Global Warming Challenged

Cost Optimized Edition

This edition is the same as the regular edition except the text and other formatting has been reduced in size to reduce the cost of printing and selling the book. (And four minor spelling/formatting errors that were missed in proofreading were fixed, the page numbers and table of contents changed, of course.) The regular edition was mostly 12pt text and this one is mostly 9pt. If you have the regular edition (copyright 2008 & 2009), <u>don't bother buying this book</u>. It's the same information.

William Hunt
M.S. Environmental Education (Conservation Biology), Southern Oregon University
B.S. Geology, Southern Oregon University
B.S. Civil Engineering Technology, Oregon Institute of Technology
Copyright 2008, 2009
All Rights Reserved

Global Warming, Challenged

William Hunt

Table of Contents

Foreword

There's A Fox in the Chicken Coop – Global Warming, Challenged
by William Hunt

"Foxes— grey, red and kit— are great for rodent control, but the red fox— Vulpes vulpes— shouldn't be guarding the chicken coop!"- W.H"

"The world's oceans are warming!" "The polar ice caps are melting!" Agriculture will fail in the face of drought, rising temperatures, increased pestilence, superstorms and hurricanes. Mass starvation will result as fisheries collapse from warming. Pikas will disappear. And so on. We've all heard that drumbeat in the daily media with headlines like that. Pressure is mounting worldwide. International leaders, U. S. politicians, political parties are declaring that man is causing the world's climate to change for the worse— that the world is "warming up", with catastrophic consequences. Mankind, if left unfettered, uncontrolled, will be responsible for the wholesale destruction of our world ecosystems. This is Global Warming, the Climate Change crisis that we hear about everyday, everywhere we go.

The ordinary citizen sees politicians, college professors, Eco-groups and the UN caught up in the media frenzy. As an engineer, scientist and conservationist, I am compelled to show an alternative side to the issue: The side of the data. The side of cold logic, such as 5+5=10, rather than -25.

I take great exception to the media's "popular" views of gloom and despair. Our future is filled with **hope**. Hope for the future.

For example, when was the last time a friend of yours died from tetanus? Or a broken leg? Or an infection from a small cut? Or smallpox? Or measles? What city in the U.S. has unsafe drinking water? Every problem that has cropped up over the years has had solutions. New problems have come up, but those, too, will have solutions. We should be cautious, not driven by emotion or panic. No sensible person should run off half-cocked— that goes for the conservationist as well as the drug-using Eco-nuts. Leave the panic to the feeble-minded. "Chicken Little" didn't gain much with his panic and neither will we. Mindless fear distracts us from real problems.

Worldwide Hysteria

In 2007, the United Nation's (IPCC) Intergovernmental Panel on Climate Change declared as an "established fact" that the Earth's climate is being warmed and that carbon dioxide liberated as the result of man's use of fossil fuels both personally and commercially is the cause. The politicos of the IPCC released press statements claiming that practically the entire world's scientific community was in agreement on the foregone climatic world-in-crisis: all nations must accept the matter and adopt the UN proposals to legislate ways to stop climate change "before it is too late." Remember that phrase, "before it is too late." You'll see it a lot in the media in stories about Global Warming and Climate Change.

Global Warming prophet and archarchon Al Gore's 2007 Oscar Award-winning movie, "An Inconvenient Truth", was seen by many to be a propaganda film purporting the same thing. To Gore and his cult's followers, climate change is a fact and global warming IS happening and it IS man's fault. That this flies in the face of what is known about atmospheric, geologic and hydrologic processes doesn't appear to matter to him or those of his religion.

An Inconvenient Truth (2006) claims that we have less than ten years left (seven now) to avoid destroying the world via weather phenomenon of Biblical proportions. It would be interesting to see how the Algorites came about this exacting time line when our weather is unpredictable past a week or two, but I digress. Gore and his "circuit-riding global warming preachers," are adamant about getting the "truth" of what he calls our "planetary emergency" out to ordinary citizens before "it's too late..."

Is there any wonder that with the media running news stories that support Global Warming and ignoring news items to the contrary, that organizations like the IPCC and Gore received the 2007 Nobel Peace Prize?

Are people like Gore heroes? Are the Greens, the IPCC, political parties and NGOs (Non-Governmental Organizations, which is UN-speak for everything from legitimate medical relief organizations to what are often lawbreaking, tetrahydrocannabinol-powered pests and occasionally even eco-terrorist groups like Animal Liberation Front) heroic in their desire to "save the world"? Are they "saving the planet from destruction"? Are they correct in their analysis and in their conclusions? Or are they something less benevolent? Is it really just a campaign to control the most basic activities of the free world?

There are foxes in our chicken coop. Red ones.

Is there Really a Scientific Consensus?

The claimed scientific consensus of the IPCC, Gore and most of the western press is the consensus of the schoolyard bully— agree with us or else. The "else" is ridicule, being called "irresponsible", an "Earth-hater" or worse. One can be discriminated against or even fired from teaching or public sector jobs. If one looks deeper than the popular media does, it can be demonstrated that a growing number of climatologists, meteorologists, geologists and other atmospheric and earth scientists DO NOT accept this global warming myth. The idea that man-produced carbon dioxide can change the climate as they claim violates several scientific principles and the data accumulated over the last century and a half. The historic record doesn't support their arguments, either.

I was astounded to find that some **31,000+** concerned scientists of different disciplines worldwide have signed a petition in La Jolla, California that essentially states that man is not causing warming of the climate by his activities. These scientists utterly reject Kyoto and other measures designed to restrict the use of energy. http://www.oism.org/pproject/

These scientists, like me, are concerned about the harm individuals and nations are by facing by the imposition of arbitrary global warming-induced restrictions and fear-based legislation. Many of these scientists further reject the idea that man even has the ability to change the climate through greenhouse gases. They are correct in this. At least a half dozen natural processes produce carbon dioxide in amounts so large as to dwarf anything man can do or would ever be able to do. It is sheer arrogance that so many have convinced themselves that man can warm the planet up, when we cannot even change the weather by one degree. Often, we can't even forecast the weather accurately, only one day in advance.

Only after almost 20 years of the building frenzy in the media have many scientists finally begun to fight the climate change claims. It's ten years past the time to do so, but any effort is hopeful. While the media has been largely uninterested in studies and reports disproving global warming, the more Climate Change is challenged, the more likely that it will finally lose its momentum and sanity will be restored for a time. At least, until the next apocalyptic craze comes along. Perhaps the next disaster will be "too much helium removed from our natural gas wells" being made into a threat to mankind or something similarly silly. Some have even tried to make nitrogen a threat.

Global Warming is actually part of a trend. In the last decade, I've seen most environmental groups go off the deep end, advocating policies that end up harming our local environments in order to somehow aid the "world's ecosphere". Lofty-sounding nonsense that ultimately ends up causing harm to the very environments that they claim to cherish.

Weather Channel founder John Coleman has been working to bring views opposing to Global Warming into the public view and his efforts have been mostly ignored. In 2008, he sued former VP Gore for fraud, in order to bring the matter to court, where the matter can be debated and evidence presented from both sides. I wish him well and hope he can accomplish what the media doesn't have the honesty to do.

A 2007 report by U.S. Senator James Inhofe of the Senate Environment Committee cited 400 scientists that disputed the claims and assumptions by Gore and his fellows. Some of the European scientists cited were actually on the IPCC panel, but were ignored by the IPCC when they didn't approve of or agree with the IPCC's determination to blame Global Warming on man's actions, regardless of studies or data. This bias alone should render the IPCC's findings questionable, at the very least.

Why is there such alarm? IS there a 'clock-ticking'? What are the true scientific facts and what can we as concerned citizens do to make our planet a better and safer place to live?

There is a clock ticking. It is not a clock for the planet, however. The time for Global Warming and its advocates is running out. The backlash has begun— with more and more scientists and politicians finally saying "enough", it has limited time.

The fox in the chicken coop is nervous because the lights in the farmer's house just turned on.

What *is* Global Warming?

Global Warming is the idea that the world has warmed since the beginning of the Industrial Age and that this warming is caused by man's increasing the levels of carbon dioxide (CO_2) in the atmosphere by his industries and other activities.

Climate Change is almost synonymous with Global Warming as it is the assumption that the world's climate is changing rapidly and that man's carbon dioxide emissions are to blame.

The primary questions this book tries to answer are:

Question 1: "Is the World's Climate Warming?"

Question 2: "Is Man Capable of Changing the Global Climate?"

When I set out to answer these questions, I had no idea that so little research and so few studies had been done in the key areas of oxygen sources, carbon dioxide sources and carbon sinks. Like most scientists, I assumed that the major universities and the various government agencies such as the Environmental Protection Agency and United States Geological Survey had performed the studies necessary to at least partially answer these questions. With most television and radio programs, and newspapers, and many politicians worldwide now attributing hurricanes, unseasonably hot or cold weather, droughts and floods to Global Warming, surely there had been at least some solid science behind the idea.

Unfortunately not.

As a lifelong conservationist I think in terms of the stewardship of natural resources and conserving species and other elements of our environment in a logical fashion, rather than elevating nature for its own sake to a form of 'nature worship' as many extreme environmentalists have done. Indeed, the first job of man was to tend the Garden of Eden. This is the stewardship role of man— to take care of what one uses.

This book asks many questions about Global Warming and Climate Change. I approached the research of this book in the spirit of "well, if the science shows that it's actually happening, I'll accept it, bitter though the knowledge might be." Wherever the evidence pointed was the view I would adopt. However, there is a severe dearth of solid information, a lack of serious studies, and a number of poorly done "studies" that would not pass peer review. Or worse, have passed peer review, despite being so badly done that no one should accept them. What information we do possess directly contradicts the claim that Global Warming/ Climate Change results from man's actions.

This book was intended to be a rigorous interrogation of the facts and figures. I spent months going through various agencies' databases, looking through peer-reviewed articles and other sources, both U.S. and international. The facts and figures that I could dig out of the National Astronautics and Space Administration, the National Oceanic and Atmospheric Administration, the National Solar Observatory, the U.S. Department of Energy, the U.S. Geological Survey, and several other agencies and other institutions are contained herein. Originally, the research on this book was exclusively science-related, but as the project developed, it was necessary to add a section dealing with the politics of Global Warming and to add a section on logical conservation.

All things considered, I've been frustrated by the lack of basic data in several areas and cannot help but ask, "Why haven't these studies been done? Why is this information not available, despite all the urgent claims by Global Warming/ Climate Change crisis apologists?"

It amazes me that a matter so blatantly political rather than scientific has made such a splash with both Congress and the Hollywood set. Former Senator Al Gore takes the Global Warming and its associated Climate Change to new levels, cherrypicking his science to "prove" that it is happening in his movie. Are movie stars and political figures jumping on a fiery bandwagon of doom and gloom to "ride the wave" of a popular cause to "save the planet" in order to further their careers? It appears that this is the case. After all, name one Hollywood type or Senator that lives in a small house, drives a small car (that is their only vehicle), and uses less total energy than the average American in a given year? After all, that is what they tell the rest of us to do. Slightly hypocritical. (Especially as many Hollywood types now claim to be Buddhists, who disdain worldly goods.)

When Global Warming and Climate Change are tossed upon the compost pile of "scientific" hysterias of the past— the comet scares of the early 1900s, the "theory" of eugenics of the first half of the 1900s, the world blowing up from nukes in the 1960s and 1970s, a new global ice age in the 1970s, Y2K in the late 1990s and many other scares— one wonders how quickly those same politicos and celebrities will disavow anything to do with Global Warming. Or take up another worthless cause. "Save the Arthropods!" or somesuch.

Rather than appeal to the emotions, this book will attempt to analyze the facts as they are.

Let the facts lead us where they will.

Section 1 The Science of Climate "Change"

1.0 Chapter 1
Is the World's Climate Warming?

The short answer to that is: we don't know with certainty. Climates are measured in terms of decades or centuries. Temperature records are spotty earlier than 1900 or so and the records since 1900 state that average temperatures have either heated up or cooled off, depending upon location. It appears to have cooled in the last decade, despite the hoopla from the Greens, the media and the Algorites, but this is not an indication of Climate Change, either. The information that we have, and more importantly, _don't have_, makes it impossible for us to know with any degree of certainty.

Accurate temperature records have existed for less than 100 years. Or perhaps the term should be "standardized." We'll get into that later. We have some wonderful methods for obtaining weather and climate data, but keep in mind that most of the technologies that we use for studying the weather didn't exist until recently. Table 1 gives an idea of how the accuracy of the measuring of temperature has changed in the last 250 years:

Table 1 Temperature Accuracy

Approximate Period	Typical Laboratory Accuracy
Prior to 1750	No Standardization, Accuracy Very Poor
1750 to 1850	Many Standards, More than 1 degree F
1850 to 1900	More than 1 degree F
1900-1950s	1 degree F
1960s, 1970s	0.1 degree F
1980s to Present	Less than 0.01 degree F

To give an example: I worked for the Corps of Engineers Materials Testing Laboratory in the Portland, Oregon area, and it was an education in just how far science and engineering had come from in just 50 years. The equipment was an interesting mix of new and old— the lab was constructed in 1947— and it was immediately noticeable how modern testing equipment was both more precise and accurate than WWII era equipment.

The advancement in electronics in that time was staggering— from vacuum tubes to transistors, integrated circuits to microprocessors, and similar improvements in thermometers, pressure gauges, strain gauges, and a multitude of other devices. There was a vast difference between using a 1950 steel test apparatus that took 24 hours to determine the density of soil versus a modern nuclear densometer that could determine the same characteristics in a few seconds.

Table 1 shows the approximate accuracy of laboratory equipment. Field equipment has also advanced rapidly as well, but due to the limitations of portability, sun, wind, rain, and other factors, almost always will be less accurate than the laboratory.

1.1 Development of Weather Instrumentation

Prior to the 1500s there was little interest and little advancement in the sciences. The Renaissance changed that with a renewed interest in learning. The 1600s saw the initial development of chemistry and modern mathematics. During the 1700s the universities and governments of several nations in Europe and North America began taking a more organized approach to the sciences. Unfortunately, the tools available then were quite crude by modern standards. Our current temperature scales of Fahrenheit and Celsius were created during this period, but there was no standardization. Of the institutions that performed scientific studies, each had their own standards for measurement. There was no way to know with certainty if a Celsius degree at Lyon, France was the same as a Celsius degree in Paris— even if the manufacturer of the thermometer was the same! Chances were that they were different by some percentage.

Later, during the 1800s, most industrial nations began to adopt standards, but most of these were still incomplete or inaccurate by modern standards. For example, a barometer or thermometer manufactured in New Jersey would be somewhat different than one manufactured in New York and thus would give different readings under the same conditions.

By the end of the 1800s, most western nations had standards for temperature and pressure, but even these often differed from nation to nation. The United Kingdom's 32-degree standard might be somewhat warmer or cooler than the United States' standard of the time, for example. There were even attempts to have some standardization of units such as the Treaty of the Meter in 1875. Unfortunately, absolute, precise, verifiable standards for measurement were years away.

During the 1900s, many countries determined methods to make absolute standards (such as the Standard Meter bar constructed of platinum and iridium in Paris) for measuring temperature, length, weight, mass, force, energy, and other measures and signed treaties to that effect. This made it possible for someone taking a measurement in one nation to be able to accurately and directly correlate that same measurement in another country, such as a scientist in Germany performing a given experiment and a scientist in Australia duplicating his test results when doing the same experiment. A degree Celsius was now a degree Celsius everywhere. The same with Fahrenheit, Rankin, and Kelvin temperature scales.

While there was interest in weather mechanics during the late 1800s and the measurement of air pressure and temperature was a hobby for many individuals and a study for a few universities, the study of weather didn't become a true science until early in the 1900s. The study of weather became systematized during the 1900s, changing from a few weather stations with manual instruments to Doppler radar and complex computer systems collating weather data.

The 1930s and 40s were a period in which the study of weather became important, with many countries, universities and scientific groups actively taking measurements and sharing information among themselves. The Second World War was the first war in which nations such as the United States and Germany actively studied the weather and used weather predictions in planning military actions. In fact, the last World War II Axis military unit to surrender was a German weather science unit on the island of Spitzbergen.

Weather was studied intensively during the 20th Century, with satellites and worldwide networks of Doppler radar eventually providing real-time data. It needs to be kept in mind that these are very recent developments, however, as shown in Table 2.

Table 2 Technological Developments in Weather

Technology	Who	Year First Used/Done
First Temperature-Measuring Devices	Galileo	1596
First Barometer	Torricelli	1643
First Organized Weather Reporting	Smithsonian	1849
First Weather Charts	U.S. Government	1869
U.S. Weather Bureau Created by Congress	U.S. Government	1870
First Daily Weather Map (of Washington D.C.)	U.S. Government	1895
First Radio Weather Broadcast	U.S. Government	1921
Weather Radiosonde (Balloon) Program Begun	U.S. Government	1937
First Tornado Warnings	U.S. Government	1948
First Weather Radar Network (WW2 technology)	U.S. Government	1957
First Weather Satellite (Television/infrared)	U.S. Government	1960
First Weather Instrument Ocean Buoy Network	U.S. Government	1970
First Effective Weather Radar Network	U.S. Government	1974

Table 2 Technological Developments in Weather, Continued

Technology	Who	Year First Used/Done
First Doppler Weather Radar Network	U.S. Government	1988

Sources: National Weather Service, U.S. Dept. of Energy, NASA

Note that the first satellite to study weather was launched in 1960. TIROS I took the first television images of Earth from satellite. Primitive and low resolution by modern standards, it was the first time that weather scientists could look at weather photos of the Earth from day to day. TIROS I was the first of many weather satellites providing data to help analyze the weather since the early 1960s. The technology has improved dramatically in the last decade or two and the precision and accuracy of the measurements have improved as well. TIROS could show a picture of a cloud mass and give an idea of the temperature. Modern satellites can image huge areas with high resolution, providing temperature, water vapor, air pressure, and other data. Today, networks of satellites now cover the entire world, providing real-time data.

To recap, the problem with determining whether Global Warming is happening is that we have data for such a short period of time. Temperature records with accurate equipment are less than 100 years old. This makes it impossible to know if the climate has warmed recently. Climatic cycles appear to be decades to centuries long and trends like the Little Ice Age or a warming period often take decades or centuries to take shape. The historical and archaeological record provides better clues for determining whether the world's climate is warming.

Another factor is that temperature stations change locations. For example, Bend, Oregon's official weather station has moved 3 times in the last 30 years and moved at least once before that. At one point, it was located in a canyon that was warmer than most of town. There was enough difference that one couldn't easily compare that location's temperatures with the other locations' temperature readings. Even a small location change can make a difference.

Several years ago, I read about a small Arizona town that moved their temperature station from the watered courthouse lawn to a dry location across the road. The official high temperatures after the move were often more than ten degrees higher than before, making the town one of the hottest in the United States. Prior to the move, the town was average for Arizona. What happened? The transpiration from the lawn grass and the watering of the grass caused a reduction of the local temperature from evaporative cooling, lowering the temperatures measured at the station. (Swamp coolers use the same principle of evaporative cooling to cool homes.)

It is a symptom of our limited time frame for data and lack of understanding of weather mechanisms that every climatic computer model that indicated a temperature increase prior to 2000 was proven to be incorrect. Most models predicted a three to six degree increase in the world temperatures by the year 2000. It didn't happen.

Every model that projected temperature increases in the early 2000s has similarly failed.

It is currently impossible to adequately model the literally millions of variables that a comprehensive global weather model would require. No computer system now in existence can handle that many variables and most of those variables are not known in detail. To make accurate predictions, at a minimum, one would have to model the winds over every landscape and body of water on the planet, taking local variables into account, as well as the amount of sunlight received, heat radiated to space, and the solar wind. Imagine the literally trillions of data points that would be needed! Were we able to calculate weather accurately by model, we could accurately predict the weather weeks in advance. As of yet, it is often impossible to predict the weather for a city or county more than a day or two in advance with certainty.

To further compound matters, only the industrial nations of Europe and the U.S. have had anything close to a comprehensive coverage of temperatures and other weather data over their landmasses over the last 100 years. Some nations, such as Canada or Australia, had weather stations in a handful of cities prior to fifty years ago, but data for other areas is nonexistent or is not consistent. Even today, large areas of Asia, Africa, Antarctica, South America and Central America have few or no weather stations, leaving tremendous gaps in our coverage of the world's climate.

With all this in mind, it's clear that a Global Warming trend would be undetectable with the temperature data that we possess and the methods we are using. In a nutshell, we can't prove the climate is changing with our current data, but we **can** prove that man's activities are not causing Climate Change.

2.0 Chapter 2
Natural Disasters and Climate Change, Oh My!
A Walk Through the Global Warming Land of Oz

Floods, hurricanes, droughts and other natural disasters have frequently been used to illustrate the "fact" of "Global Warming." Enter the witch on the broom, Al Gore, and his horde of winged monkeys selling global warming. Claims are made almost daily that an event or condition is a symptom of Climate Change. Winter storms, summer heatwaves, and even hurricanes are said to be the result of Global Warming. If temperatures are hot, cold, above average, below average, or average, there are those that will say that it is the result of climate change. If there is a drought or more precipitation than the average, it's climate change. And so on and so on, ad nauseum.

None of this is logical, as all of these events have been common for as long as man has lived on the planet. Even earthquakes have been occasionally blamed on climate change by the Global Warming lobby, despite earthquakes not having a connection to weather. The worst storm, flood, and drought disasters ever recorded in the U.S., in terms of lost lives, all occurred before 1930, prior to Al Gore's entry into the world.

The Johnstown Flood was the worst in recorded U.S. history, with 2200 deaths resulting from a dam failure in 1889. The worst hurricane was the 1900 Galveston Hurricane that killed as many as 8000. The worst tornado in U.S. history was the "Tri-State Tornado" in 1925, killing 689 people in three states.

The most severe earthquake the U.S. has ever experienced was a 9.2 in Alaska in 1962. The most severe quakes in the lower 48 states are three quakes estimated to be roughly an 8 on the moment-magnitude scale in the New Madrid area of Missouri, Tennessee, and Arkansas in 1811 and 1812.

There are three reasons why natural disasters figure so prominently in the last thirty years:

1. Modern media highlights each and every hurricane, earthquake, and flood as if it were unusual. How many times have we heard "worst in a hundred years" about a storm or hurricane, when a couple of decades before, some other disaster dwarfed the current storm? By their logic of "100 year storms," we've literally had back-to-back hundred year storms in some areas, which says more about the way we determine the probability of major storm events than the storms themselves. It sells airtime and newspapers to make a disaster sound worse than it is.

The last year I worked in northern California, there was a 100 year storm. Sounds like a major event, right? It affected only one tiny creek, due to a cloudburst localized over that creek's watershed. Had the media noticed that event, they likely would have blown it up with headlines screaming "100 Year Storm in California!" or something similar and added text about it being a symptom of Climate Change. It's all in how it's portrayed in the media. Al or Pelosi could have used it to "prove" that Global Warming is upon us, through the wonders of the media.

How many times have we heard of "the Big One" being predicted for San Francisco? We had another "Big One" in 1989, with the Loma Prieta earthquake. The 1906 SF quake has been estimated at 7.7 on the Moment-Magnitude Scale— the Richter Scale hasn't been used since the 1980s— and the Loma Prieta was a 7.1. While there will likely be larger quakes at some point in the future, the recent hype of a repeat of the Great Quake of 1906 is a little silly, given that an event almost as large occurred in 1989 and the results were not nearly as serious as 1906. With today's modern construction and emergency systems, it would require a much, much larger quake than the 1906 quake to cause damage like that seen in 1906.

Threes and fours occur daily in California and Alaska and it never makes the media unless it is a slow day. If a 9.5 struck Southern California, it would kill thousands and cripple the transportation system. The 6.7 Northridge Quake of 1994 would have killed thousands had it happened fifty years earlier. As it was, life quickly returned to normal with a minimum of lives lost.

I've yet to hear the claim of a link between climate change and the frequency of earthquakes, but I'm certain someone will invent one. How about a nonsensical theory like "warming causes the crust to heat, causing expansion, resulting in earthquakes" or something similarly impossible? Governor Kulongoski of Oregon will figure something out, no doubt, but I digress…

2. Secondly, there are many more people living in harm's way than in years past. Millions now live in coastal areas that are vulnerable to hurricanes, typhoons, and cyclones. More than 30 million live in areas of California that are at risk of a severe earthquake. Hundreds of thousands of people now live on subdivisions built on Atlantic barrier islands whose sand moves every time there is a major storm.

3. Lastly, man has removed many of the buffers between himself and potential natural disasters. Worldwide, mangrove swamps and other buffers against the ocean have been cut down for wood and other needs. Mangrove swamps dramatically reduce the force of incoming storms, protecting coastal lands. New Orleans had lost most of its buffering swamps to development and with its explosion in population living below sealevel, was ripe for a disaster like Katrina.

2.1 Floods

The most common natural disaster is the flood. Most of the world's rivers have seasonal high and low flows. All experience extremes at different times of high and low water levels. The worst flood on record was the 1931 Yellow River Flood, which killed between two and four million Chinese.

Forests in Europe and Asia have undergone drastic changes over the centuries and cover only a tiny fraction of their prior range. The conversion of forest to croplands, grasslands, and scrublands, and especially to urban landscapes of stone, concrete, and asphalt, allows runoff from rainfall and snowmelt to collect and fill rivers much more rapidly than historically.

In Europe, many streams have been channeled by concrete and stone banks, dikes, and dams. The Netherlands has worked mightily to prevent flooding by creating a tremendous water management system. For the most part, these work well. It is generally only the largest storm events that now trouble Europeans. However, when a major storm happens, it now is claimed to be the result of Climate Change or at least exacerbated by Climate Change due to Global Warming—even though such storms are well within the normal range of weather extremes. The North Sea wasn't called the Poison Sea by its neighbors because of its gentle nature.

The U.S. has more than 11,000 dams, as well as dikes, levees and other water channeling/impounding structures. Most, like the dam systems on the Columbia River, the Missouri/Mississippi Rivers, and the Colorado River, work very well to prevent seasonal flooding. Major flood events, such as the "Great Flood of 1993" and the Columbus Day Storm of 1962, wreaked havoc in the past, and events like that, though rare in most of the nation, will happen again in the future. For more than a decade, however, the Global Warming lobby has had a field day every time a river floods, calling each event "symptomatic" or "proof" of Climate Change caused by Global Warming, with innumerable televised "talking heads" of sagely mien, wagging tongues and vacuous craniums.

Most nations of the world do not have adequate flood control and many have large populations at risk. Asia is much more haphazard in their flood management than Europe or the Americas. Some areas have dams and other flood control structures, while many have literally nothing between their populations and floods.

Bangladesh is one of the worst examples of people living in unsafe areas, with tens of millions living in low-lying areas that flood both from the annual monsoons and from every cyclone that moves in from the Indian Ocean. Nearly every year, thousands are killed— swept away in floods or from the diseases caused by flooding.

The Brahmaputra River has seen devastating floods in recent years. This has been blamed on Global Warming. Flooding of the Brahmaputra is not unusual, but the intensity has worsened in recent years because the forestlands and riparian areas of Tibet, Nepal, Bhutan, China, and India that would have served as buffers against floodwaters have been damaged or completely removed. The tributaries of the river and the main river itself pass through lands where the buffering vegetation has been dramatically reduced or completely removed. When the monsoon or a storm drenches the land, there are fewer trees to break up the force of the incoming rain and the vegetation that would normally slow the water down is reduced, changed or completely gone. The rainwater collects and moves downstream much more rapidly than would have been the case before the forests and other vegetation was removed or changed. The result is flooding downstream in India that is much worse than in years past. This is also blamed on Global Warming. Most nations so affected go along with the lie because it's much easier to blame the wealthy countries instead of one's own agricultural and forest practices than to fix the problem.

Tillamook, on the Oregon coast, makes the national news almost every year for flooding. One could question the wisdom of placing a city in the common floodplain of FOUR rivers and numerous smaller streams. Every year the winter rains cause some flooding and sometimes more than once per year. Over the years, cattle have been threatened by floodwaters, so piles of rock have been built up on many farmers' pastures to give livestock a place to escape to. However, most home and business owners living in the flood plain have done little or nothing to protect their structures from flooding. While it's unusual for people to die in Tillamook's perennial floods, property damage is a given, every year.

There have been more than a dozen major floods in Texas during my lifetime. Every two or three years, heavy rains or hurricanes cause homes and businesses to wash away. Despite this, many people live in floodplains with no thought to the occasional flood.

Ashland, Oregon rests on an outlying ridge of the Siskiyou Mountains, where one would not think it to be a place at risk of floods. However, Ashland made the national news in 1997 when 17 inches of rain fell in two days in the eastern Siskiyous of southern Oregon and northern California. Half of Lithia Park, which is in a steeply inclined trough cut by Ashland Creek, washed away and the downtown area was flooded as the creek left its banks. The main business district was built on the floodplain of the creek and the creek itself had been locked into a narrow channel. The basements of these businesses faired particularly badly in the flood. To make matters worse, the main bridge into Lithia Park, just upstream from the business district, was sunken down into the creek channel, blocking roughly half of the channel's capacity.

Interestingly, prior to the '97 event, one of my geology professors at the local university had been showing photos of the flooded-out downtown area from 20+ years before and had commented that it was just a matter of time until it happened again. Despite the history of floods, I would not be surprised if the local city council, known for keeping the 1960s alive, along with its profligate drug use, blamed it on Global Warming. Again, though, the floods were nothing unusual. Every 20 to 30 years, this area experienced precisely this type of flooding. Many of the floods, especially, the 1962 flood, were much more severe.

The Midwest had a "sixth great lake" during the "Great Flood of 1993". The flood was accentuated, oddly enough, by the system of dikes and dams in the Missouri/Mississippi system. Normal floods, such as those from spring melt, are usually contained without problems by this system. However this storm event was so large and prolonged and the soils so saturated from winter rains that it overloaded the water management system. There were no safe places to store such incredible amounts of water and it was coming in too fast to get rid of it quickly enough to avoid property damage. The drainage system was completely overloaded. The Corps of Engineers was in the unenviable position of determining where to direct the water so that it would cause the least amount of damage. After the waters finally receded, the damage tally was $15 billion with more than 10,000 homes wiped out and another 40,000 damaged. In such a large flood event, it's not possible to entirely protect the public's private property. The only sure way to avoid any possibility of such property damage from happening again would be to not allow people to live in the floodplain, This is impractical, given that twenty million acres were affected in nine states. A better option would be to build homes and other buildings with flooding in mind so as to minimize damage. The government bought up a number of homes in order to prevent them from flooding again, but this was only a small percentage of the homes that would be flooded with a similar event.

Today, any flood event like those described above would be blamed on Global Warming/Climate Change, despite the history of such events. Why? Because the media has a short memory and it pleases them to support the ideas of Global Warming and Climate Change.

2.2 Hurricanes

Hurricanes, also called typhoons (Pacific Ocean) and cyclones (Indian Ocean), are a yearly occurrence, but do not generally strike the same areas in a given year. Hurricanes form almost every year, but usually do not make landfall in the United States. The 2007 Hurricane Season was very unusual as no Atlantic tropical storms reached hurricane windspeeds (73+ mph) and so none were of hurricane status. The number and frequency of hurricanes reported has increased over the last century, but this is the result of better reporting as most hurricanes sixty or more years ago went unreported unless they made landfall or passed through or near a shipping lane where a ship would encounter it or could see it in the distance. The ability to track hurricanes hundreds of miles out to sea didn't exist prior to satellites and weather radar.

Hurricanes can be terrifying. With sustained winds of 73 to 156 miles per hour, they can wreak havoc, blowing down trees, damaging buildings, and destroying crops. In numbers of casualties, all of the worst hurricanes of the past two centuries have occurred outside of the United States. The most terrible hurricane in recent centuries was in Bangladesh in 1970 with a million dead.

Hurricanes produce a "storm surge" which can raise sea levels locally by several feet during landfall and may dump several inches of rain on the coast where landfall is made. Depending upon on local conditions, the extra rain may be a blessing or a curse, both for surface water and groundwater. Hurricanes can also result in storms inland.

Some theories hold that hurricanes are "fed" by warm water, such as that found in the Gulf Stream. This leads the Global Warming advocates to claim that hurricanes will increase in the future, both in intensity and in numbers as waters "warm" from Climate Change. That the oceans have not been observed to be warming, and indeed, there is nothing to compare modern oceanic temperature data to— as mentioned in the first chapter— has either eluded the Global Warming lobby or they have ignored it.

2.3 Droughts

Droughts can happen anywhere. Regional droughts are normal occurrences and are common in the world's deserts and semi-arid areas. The western United States frequently has cycles of drought, with several years of reduced rainfall in the 1930s, 1940s, late 1970s and 1980s. More unusual, the early 2000s saw considerable droughts in the eastern half of the U.S. as well.

The 20th century was one of the three wettest centuries for the U.S. of the last eleven, according to a number of tree ring studies. A severe, prolonged drought period occurred in Mexico and parts of the U.S. from 800 A.D. or so through 1100 A.D. that may well have ended Mayan and other societies. The American Southwest experienced an extreme drought from 1000 A.D. to 1300 A.D. There are indications that there may have been a 3-year span entirely without rain around 1000 A.D.

In terms of climate, shorter-term droughts of a decade or two do not indicate warming or cooling, as they appear to be cyclical. However, a worldwide, prolonged drought that lasts several decades would indicate a reduction in global temperatures. With global cooling, the oceans' temperatures are reduced and with them, the evaporation rates and resulting rainfall. In a nutshell, cooler ocean water would mean less rainfall globally.

2.4 Tornadoes

Tornadoes happen worldwide, but are most common in the United States' "Tornado Alley," which runs from North Dakota through Texas. Some localities, like the Ozarks, have fewer tornadoes than the nearby open plains. The United States averages almost 900 tornadoes per year with some years reporting fifteen hundred or more.

Tornadoes are born in severe thunderstorms, usually from thunderstorms clusters called "super-cells". The winds in a tornado can reach 300+ miles per hour and can do immense damage. The largest single tornado on record was the aforementioned "Tri-State Tornado," which traveled hundreds of miles in a straight line and changed direction only once during its existence, leaving a swath of destruction more than a mile wide in its wake. The worst "outbreak" of tornadoes in recorded U.S. history was the "Super Tornado Outbreak" of 1974, when a stunning 148 tornadoes in only sixteen hours were recorded in 13 states from the Midwest to the Atlantic.

The Global Warming lobby is predicting that tornadoes will become much more common as temperatures "rise." It is problematic as to whether severe weather would increase if the world warmed, but it does not appear likely, given that natural disasters appear most commonly during the cooler periods in recorded history.

2.5 Wildfires: Disasters Where Man Lends a Hand

One of the continual refrains of the Global Warming lobby is that the large fires of late have been the result of climate change and that fires will become more common, widespread and severe as the climate "warms." This has no basis in fact, whatsoever.

Fires in forestlands have been with us as long as there have been forests. Lightning-caused fires have been common from the Appalachians west, from the tall grass prairies of the Midwest to the Pacific Coast. The Blue Mountains, for example, get their name from summer wildfire smoke that makes them appear blue. Human-caused fires were once the normal situation in most parts of the United States, with American Indians from the East Coast to the West Coast using fire as a management tool to improve hunting and to change the vegetation composition, i.e., encourage grass versus shrubs, or to reduce future fire danger.

The 1871 Peshtigo fire in Wisconsin was the worst fire in U.S. history, leaving 1200 to 1500 dead and many injured. The fire was the result of sloppy forest practices that left heavy slash and other debris over a huge area. When the fuels dried out during the summer, a few sparks was all it took to start a fire that burned more than 5900 square miles. Imagine how the media would portray that if it happened today!

Fires have been very common during the 20th and 21st Centuries. It varies from year to year and region to region in intensity and the number of acres burned. Some years, like 1910, in which 3 million acres burned in Idaho and Montana alone, are remembered among forestry professionals. 2008, with more than 1700 fires throughout northern California, 1680+ of which started from one massive storm system during the weekend of June 20, 2008, will also be remembered.

Fires in the last two decades have been on the rise in the western United States. There are two main causes. Deferred maintenance and fire suppression.

2.5.1 Maintenance Deferred.

Most Federal forestlands in the western U.S. – those that were not protected in some way – roadless areas, wilderness, leave strips, experimental forests, etc. – were cut for the first time from 1976 to 1987. Statistics vary, depending upon source, but approximately 96% of the non-protected Willamette National Forest's lands were cut from 1976 to 1987, for example. Most private forestlands were cut for the first time from the 1960s to 1990 or so. The Federal public lands of Oregon and Washington— mostly Forest Service and BLM— produced 11 billion board feet of timber cut in 1981 and increased to a stunning 18 billion by 1987— until the available trees for cutting began to run out and disappearing species like the Northern Spotted Owl started putting teeth into the Endangered Species Act.

Some areas of northern California were subjected to more selective cutting than the standard clear-cutting, but overall, the history in the Klamath and Sierra Mountains is similar to Oregon and Washington. Arizona, New Mexico, Colorado, Idaho, Montana, Utah and Wyoming also cut their Federal forestlands during the same period and have reaped the same problem: the western timber industry was allowed to eat itself out of house and home in less than 30 years and as a result, we have tens of millions of acres composed of densely packed small trees because most stands were not maintained after the first cut.

When the first clear-cutting of a given acre of forest is cut, typically the trees grow in rapidly and in profusion. Brush and grass also grow quickly. The result is a heavy fuel loading. After fifteen or twenty years, depending upon growth rate, the trees need to be thinned out, much like vegetables in a garden are thinned. Most of the acreage of forest cut from '76 to '87 was not thinned or otherwise maintained.

When not maintained, the trees continue to grow and cluster together, striving against each other for sunlight. The grass is out-competed for sunlight by shrubs and the shrubs take over the undergrowth. The close spacing of the trees, coupled with heavy shrub growth, makes for a very dangerous fire situation with the entire stand being replaced if fire gets started. The grass starts on fire very easily when dry, then its heat starts the small twigs on the forest floor, then the twigs spread fire to the brush, and so on. Pretty quick, the entire stand of trees is on fire and it spreads rapidly, the rate depending on how dry the fuel is, how hot it is, the humidity and how much wind. Dry wood, hot weather, dry air and a stiff breeze make for an inferno and explosive fire growth. "Like gasoline" is a phrase often used in the fire biz.

With trees packed closely together, they tend to be under stress because of competition for sunlight and water. This stress makes them more susceptible to disease and insects. The spruce budworm, *Choristoneura occidentalis*, has frequently plagued the forests of the Rockies and Pacific Northwest. The Mountain Pine Beetle, *Dendroctonus ponderosae* and multiple species of *Ips* bark and boring beetles have attacked trees throughout the west. Such vectors are a natural component of the forests, but when trees are unhealthy and packed together, it's like a buffet table for the bugs. Large areas of trees can be killed off in a single season. Dead trees burn very nicely, especially after they dry out.

Some agencies, like the Bureau of Land Management, have done extensive thinning and fuel reductions. The Forest Service has been trying to do stand maintenance and reduce fuel loadings, but has been largely hampered by Congress' reduction of personnel since 1992 and sporadic, inadequate funding for forest maintenance. Congress finally started allocating money for more forest maintenance (thinning and fuel reductions are still desperately needed in most areas) when the 2000 fire season stung them into action, but they still will not allow the agencies involved to hire the personnel needed to fix the problem. At present, some regions such as California and Washington/Oregon are staffed at 1/3 the level (of non-fire positions) that they had in 1985, and with the incredible number of lawsuits, it's almost impossible for anything to get done.

2.5.2 Fire Suppression

From the late 1800s to the 1970s, fire was vigorously suppressed. This allowed fuel to accumulate in drier areas of the forests that had been previously been removed by more frequent fires. The initial result was fewer fires, but the fires that did happen had greater intensity due to heavier fuel loadings. Eventually, however, the fire frequency increased as the forests became choked with fuels.

Compounding the fire problem is that many more thousands of people are living in the woods or on the edge of the woods–what forest planners call the "Wildland Urban Interface" or WUI (pronounced woo-eeee) for short. Now, when a fire spreads through wooded areas, it may encounter houses and other buildings. With homes in forested areas, it becomes critical to clear a "defensible space" around the houses to keep fire away from the walls and to construct buildings of fire-resistant or fireproof materials such as stone or concrete.

Given the background of our forest situation, to blame "Global Warming" and "Climate Change" for the increase in fires is not reasonable. Especially when one considers that a warming climate would cause increased rains and likely, fewer fires, because of higher humidities.

For many years disasters were called "Acts of God". Now they are effectively thought of as acts of man, resulting from the mistaken idea that man can cause them or make them more severe and frequent. The problem with that idea is clear: when was the last time man could control the weather or even the path of a single storm?

With the exception of wildfire, is not reasonable to claim increases in natural disasters like hurricanes, tornadoes, floods and other natural disasters are the result of man's actions.

3.0 Chapter 3
Is Man Capable of Changing the Global Climate?

So, if we can't know if Global Warming is happening, shouldn't we do something about it on the chance that it is happening?

Absolutely not. That question is often asked by environmentalists as a justification for pushing ahead with policies that restrict or prevent energy use in order to reduce carbon dioxide emissions— which would effectively destroy entire economies. The idea that man can affect the global climate is presumptuous at best. Natural processes **completely** dwarf man's activities and any energy that man can wield. When one compares the magnitudes of the forces and energy inherent in volcanism, solar energy, oceanic and land plants, animals and geologic processes, man's activities shrink to insignificance.

The Earth's climate is influenced by the sun's heat, solar wind, Earth's own internal heat sources, and the thermal properties of the atmosphere, "dry" land, and the land and ocean waters. The atmosphere itself is mostly the product of hydrologic-, geologic- and plantlife-based processes, including photosynthesis, carbonate dissolution, carbonate deposition, and volcanic outgassing. The oceans, fresh water bodies, atmosphere and to a lesser extent, the surface soils and rocks, serve as heat reservoirs that store and release heat.

3.1 The Earth's Heat Management

The idea behind Global Warming is that the amount of heat that is retained by the atmosphere is increased by man adding carbon dioxide to the atmosphere. As will be demonstrated, this has no basis in fact.

To indirectly determine if the world is warming, it's necessary to quantify how much heat the Earth is receiving, how much it absorbs, how much it retains and how much it releases back into space.

3.1.1 Energy from the Sun

The Earth's primary heat source is the Sun. In any given moment, the Earth is being irradiated by an average of 1365.3 watts per square meter, which is 349 trillion kilowatts spread over the surface of the Earth that is lit by sunlight. That is 3,059,334,000,000,000,000 or 3 quintillion kilowatt-hours per year.

Now, to give an idea of how minuscule man's abilities really are, let's compare that figure with man's ability to generate energy. The total world electric generation in 2005 was 17,350,575,965,208 or 17 trillion kilowatt hours.

Now, compare the two numbers.

The sun bathes the earth in roughly 176,000 times the amount of electricity that man can generate. This is like comparing the total land area of New York City to the total land area of the Earth.

Still think man has changed the climate? Or has the ability to?

The sun varies in its output considerably, roughly 5 watts per square meter more or less than the average of 1365.3 watts per square meter during the last 22 years. That is only 0.7%. Even ±5 watts of difference causes an enormous change in the total heat and light the Earth receives from the Sun. This variance of 10 watts per square meter calculates to almost 1300 times the amount of electricity that man generates. Per year, this is more than the amount of energy that man has ever used in his entire history. More energy than all the nuclear bombs and other man-made devices that have ever been, could release.

This illustrates how the sun completely outclasses anything man is able to do. Imagine how much the sun's variation alone changes our climate from month to month. See Figure 1, Heat Exchange.

The Global Warming/Climate Change lobby wants us to believe that we can change the climate. So might an ant try to change the course of a supertanker.

3.1.2 Residual Heat and Radioisotopic Heating

Heat left over from the initial formation of the planet is present in great amounts. Called by different names, this residual heat makes its way to the surface in the same way that heat from radioisotopic heating does. The liquid iron outer core and semi-solid plastic mantle of the Earth are indicative of this heat. It's impossible to know directly, but some estimates suggest that the inner core may have a temperature of 10,000 degrees or more. This is the Earth's ultimate heat reservoir.

Another, less quantifiable, heat source is heating from radioisotope decay. It's theorized that there are substantial quantities of radioactive elements in the Earth's mantle and core. When radioactive elements decay, they produce heat, light or other particles that are changed into heat by the surrounding materials. It is impossible to accurately estimate the amount of heat produced without actually being able to sample the mantle and core materials and thus no figures have been widely accepted.

Another source of heating yet is compressive heating, resulting from the compression of crustal rocks onto mantle, mantle rocks onto core. How much heat is generated this way is also unknown.

Whatever the source of the heat in the Earth's mantle and core, this heat gradually makes its way to the surface through volcanism and heat transfer. The heat from the Earth's interior influences the weather far more than man is able to, given the magnitude of past volcanic eruptions and even the heat encountered in deep mines, such as those of South Africa. The are no currently no accepted figures.

3.1.3 Heat Losses

There are two ways in which heat leaves the earth. The first is the radiation of heat from the surface and atmosphere to space. The second is for gas molecules to be lost from the uppermost atmosphere to space.

Heat is constantly lost from the atmosphere and land to space through longwave infrared radiation. This mechanism of heat loss has been felt by almost everyone on a personal level— when a person stands out in the open at night under a clear sky and feels the heat escaping from his/her body, that person is losing longwave infrared heat to space as well as losing heat to the surrounding air.

Another mechanism for heat loss is the loss of atmospheric gas to space. Each molecule of air that leaves the Earth's atmosphere takes some small amount of heat energy with it.

3.1.4 Earth's Insulation— Clouds

Clouds insulate the Earth by slowing the transmission of heat, much like the insulation in homes does. At night, clouds slow infrared from escaping to space, reducing the amount of heat that escapes, keeping the air temperatures below them warmer than on a clear night.

During the day, clouds reflect much of the sunlight the Earth receives back into space, preventing much of the sun's heat from reaching the land and ocean beneath them, keeping them cooler.

These factors for the reduction in heat transfer rates by clouds also have no widely accepted global figures.

3.1.5 Earth's Surficial Heat Storage System— Water

Bodies of water and atmospheric humidity moderate the Earth's surface and atmospheric temperatures. Water absorbs heat from the sun and this heat warms cool air masses. Oceanic currents of warm water such as the Gulf Stream and the Kuroshio Current warm Europe's and southern Japan's climates. Similarly, the California and Humboldt Currents cool the climates of western North America and western South America. Areas near or surrounded by ocean tend to have little variation in temperatures, due to the ocean's moderating effects. Arcata, California, for example, has less than 30 degrees difference in temperatures during a typical year.

Conversely, Bend, Oregon, on the edge of Oregon's High Desert, 200 miles from the Pacific Ocean, frequently has daily humidity of less than 20% during the dry half of the year. As a result, the temperatures typically vary by 40 to 50 degrees between a summer's day and night. Frost is not uncommon in July and September, much to the gardener's dismay. *Author's note: I've seen it snow twice on the 4th of July in Bend.*

Large landmasses, such as Siberia, have little water to moderate their temperatures, resulting in temperature extremes. In constructing the Baikal-Amur Railway (BAM) through southern Siberia, special steel alloys had to be used to prevent track breakage from contraction in the −40°F winter weather and from expansion in the 80 degree summer heat. An extreme example, Verkhoyansk, in northern Siberia has had temperatures as high as 102°F and as low as −94°F with normal temperatures ranging from the 70s in the summer to −50°F during the winter.

3.2 Sunspots and Climate

We know that the climate has been both hotter and colder at different times in the past, given the writings of historians and the results of archeological digs. The sun appears to be the main cause. This would seem intuitive, as the Earth receives most of its energy from the sun. Sunspots give us an indirect indicator of how much energy the sun is emitting.

The number of sunspots observed on the sun appear to be proportional to the Total Solar Irradiance (TSI), which is the total amount of energy that the Earth receives from the sun, expressed as watts per square meter. Our data for TSI only goes back ~23 years, but a trend is apparent. When the number of sunspots increases, the TSI increases, and vice versa, as shown in Figure 2, Sunspots Versus Total Solar Irradiance. In order to show the variation more effectively the variance in TSI is exaggerated by a factor of 750 with respect to sunspot numbers. As shown in Section 3.1.1, a small increase or decrease in the Sun's output makes a terrific difference in the heat the Earth receives.

Take a look at Figure 2 again. Notice that there is a noticeable drop in the TSI values in the upper right portion of the chart. In 2004, both sunspot and irradiance values dropped. This is noteworthy as it roughly corresponds to the 2004-5 winter wherein month after month of excessively cold weather east of the Rockies was recorded. **If** this is direct cause and effect, then the lesson is: less heat from the sun, lower temperatures. It will be interesting to see if there is a marked cooling over the long term or if the cold weather and TSI drop are merely coincidence.

Global temperatures have dropped since 2004 and so have sunspot numbers.

Sunspot numbers have been counted since 1610 and regularly recorded since 1650. During the Little Ice Age of 1350-1850, sunspot numbers were the lowest that have been recorded for years at a time. (See Section 4.4, the Little Ice Age). (See Figure 3, Sunspots from 1650 to 2006.) Our recent crash in sunspot numbers may be a trend or it might be a statistical blip, with numbers increasing again next year.

But what about that bugaboo of the greenhouse effect advocates, carbon dioxide?

3.3 Carbon Dioxide and Oxygen Mechanics

The claim is made by the Global Warming/Climate Change lobby that increased carbon dioxide is the culprit for "increased" temperature. This is based upon the flawed idea that carbon dioxide absorbs more heat and retains more heat than other atmospheric gases. This is not the case, as will be demonstrated shortly.

What is carbon dioxide? Carbon dioxide is a relatively inert, colorless gas that makes up 0.033% of our atmosphere. It is produced by a multitude of chemical reactions. Essentially all of the carbon dioxide in the atmosphere is the result of natural processes. Volcanism, plant and animal respiration, soil microbes (rot and other soil-forming processes as well as respiration), and the erosion and disintegration of carbonates like limestone produce carbon dioxide. Each of these processes produce more carbon dioxide than all of man's activities combined.

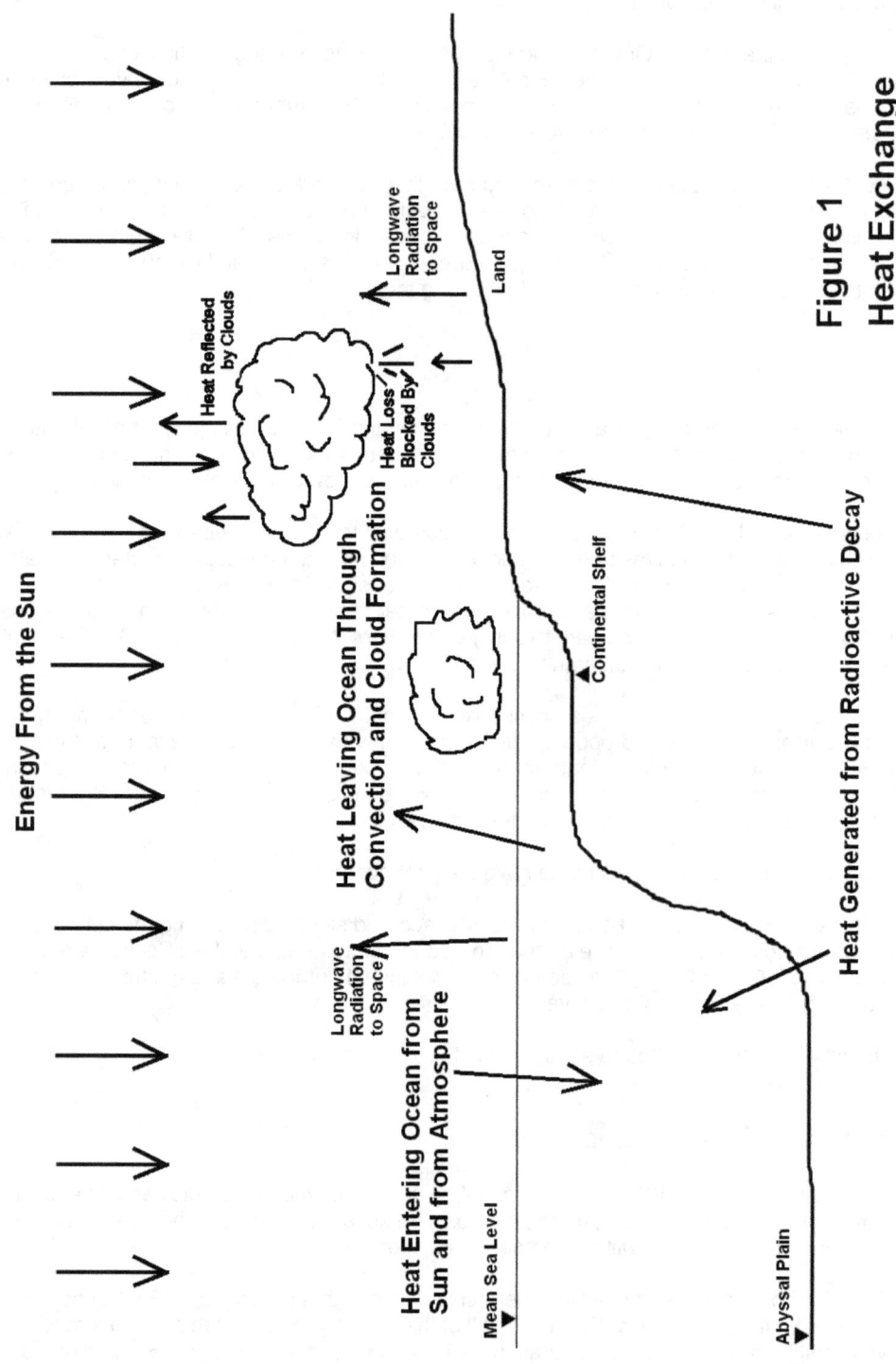

Figure 1
Heat Exchange

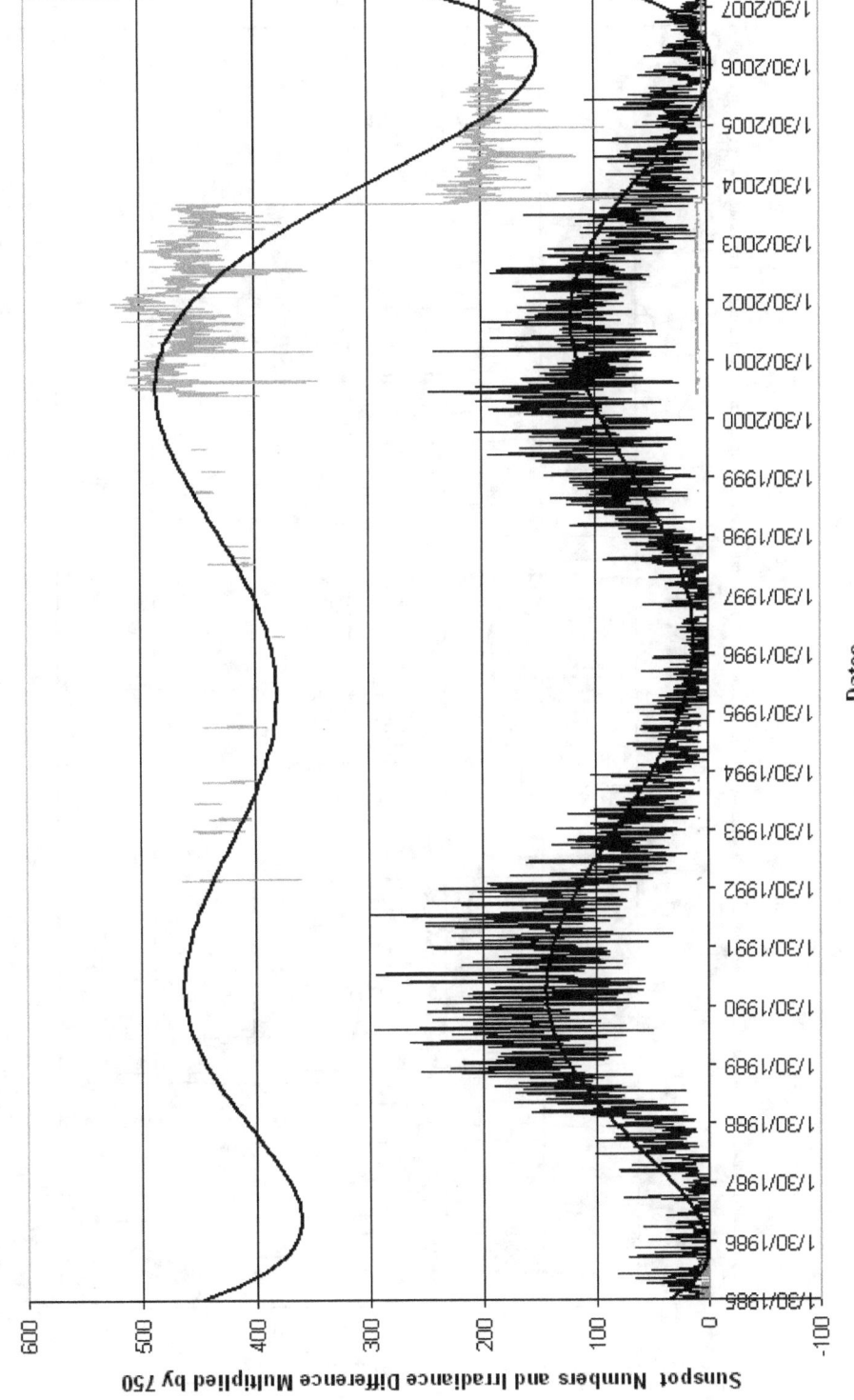

Figure 2 Sunspots versus Irradiance

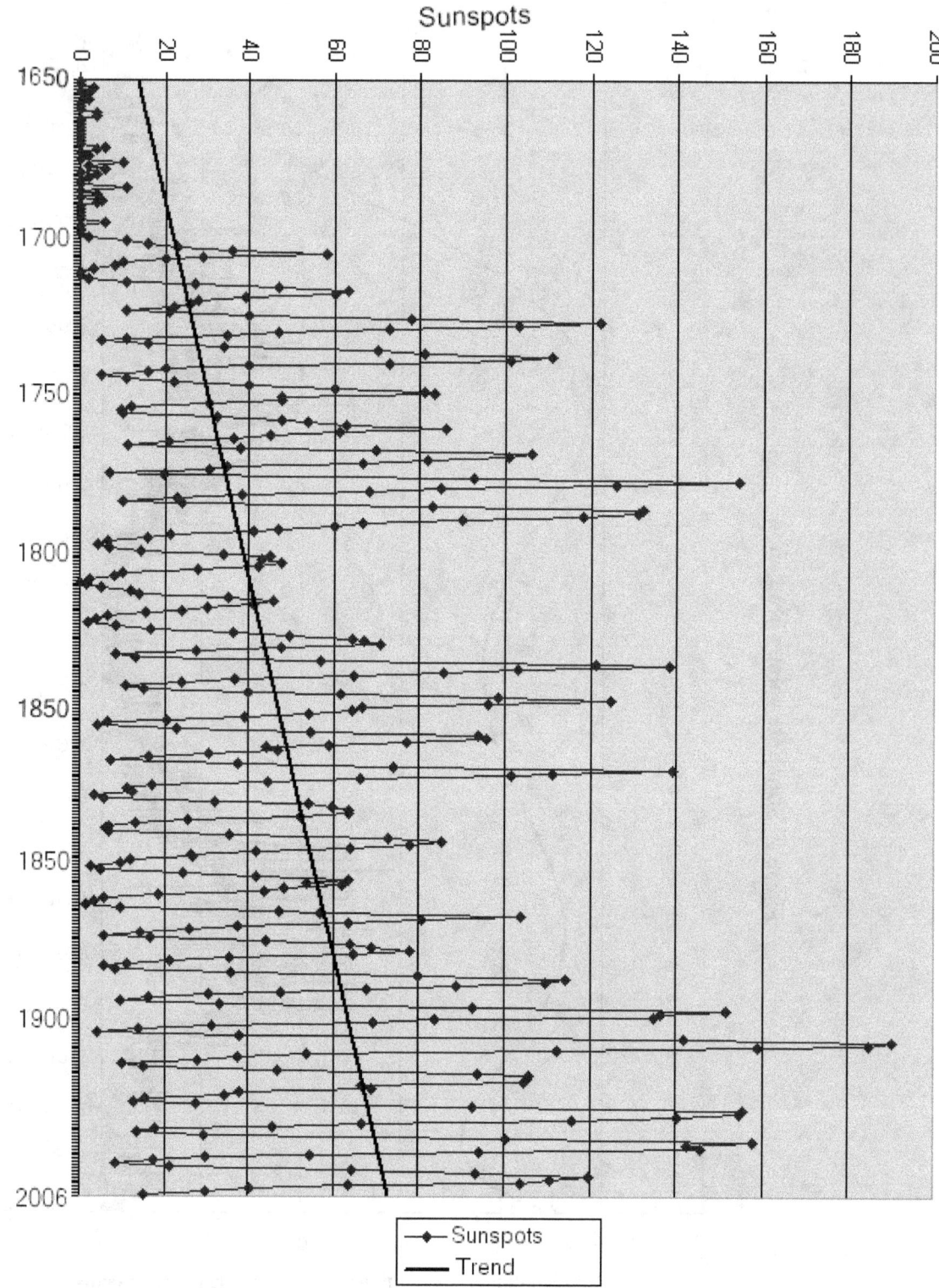

Figure 3
Sunspots from 1650 to 2006

Man produces carbon dioxide through the combustion of fossil fuels like coal, natural gas, kerosene, diesel and gasoline, the burning of wood and dung for fires and the burning off of forest and field for agriculture. Several chemical processes used in industry and commerce also produce carbon dioxide. These have been used to produce an estimate of 29 billion tons of carbon dioxide produced by man in 2006.

Man takes in food, digests it and excretes waste products that are used by microbes that further produce carbon dioxide (as well as other gases like hydrogen sulfide and ammonia). Humans respirate several times per minute, releasing carbon dioxide. These factors have not been added to any of the published estimates of the Global Warming lobby.

Carbon dioxide is absorbed into ocean water and fresh water and is also released from water over time. Bodies of water play an important role in regulating the amount of carbon dioxide in the atmosphere. Gases are exchanged between the surface layer of water and the atmosphere, speeded by wave action mixing the air with the water, especially during periods of turbulence, such as during a storm. See Figure 4, Carbon Dioxide Sources.

3.4 Carbon Dioxide Measurements

No long term studies for carbon dioxide in the atmosphere exist except that of the Scripps Institute's (later the National Oceanic and Atmospheric Administration's) sensors on the Big Island of Hawaii from 1958 to 2008. This is the carbon dioxide data that *all* environmental groups quote from. There are measurements in other locations and shorter-term studies in recent years, but nothing long term.

There is a serious error associated with the Scripps/NOAA carbon dioxide sensors. This error is so large that it makes it useless for measuring anything but local conditions, i.e., the readings can't be used for measuring global carbon dioxide levels.

The carbon dioxide levels indicated by the Mauna Loa location's sensors in 2007 were approximately 50 parts per million greater than other locations around the globe: NOAA listed an average of 330 parts per million in 2007 but the Mauna Loa data indicated 380 in 2007. This is due primarily to volcanism on the Big Island. The sensors on Mauna Loa are more influenced by the Kilauea eruptions and other nearby volcanic vents than by the world's atmosphere as a whole. Placing the carbon dioxide sensors so close to a volcano that routinely creates **VOG**– a noxious, sulfuric acid-laden **V**olcanic f**OG** resulting from volcanic gases mixing with water vapor— some *200 miles away on Oahu* would tend to skew the results, don't you think?

Author's note: I'd love to know what the recent increase in Kilauea's outgassing that closed that portion of the park did to the CO_2 sensor readings. Probably off the chart.

Having that kind of outgassing so near a carbon dioxide sensor makes any readings that sensor yields hopelessly inaccurate. It's like having a carbon dioxide sensor placed over the smokestack of a steel mill and then saying that the carbon dioxide levels in the atmosphere are very high *everywhere.*

Another possible factor in Hawaii's local carbon dioxide levels is the population explosion— Hawaii's population has doubled since 1960. Many of the fields that used to grow pineapples and other crops are now under houses. More industry, more vehicles, more garbage dumps, more construction and more of pretty much everything human would tend to increase carbon dioxide locally, but this effect would be small compared to local volcanism.

In order to make a reasonable claim that the atmospheric carbon dioxide is increasing, multiple sensors located around the world would be needed, taking readings for decades. Our data is spotty at best. The Mauna Loa sensor has done a good job of reading local conditions created by the volcanoes, but it is not reasonable to extrapolate its measurements for the world in general with such a huge source of carbon dioxide only a few miles away from the sensor.

3.5 Volcanism and Carbon Dioxide

Volcanism accounts for much of the atmosphere's carbon dioxide. The world's land and oceanic volcanic vents produce an estimated minimum of 3.65 trillion tons annually. This number is a conservative estimate— the actual amount of carbon dioxide produced from volcanism is likely to be much larger.

At any given time in the last decade, there have been roughly 500 active volcanic magma fields on land. This is a problematic number, because many volcanoes have more than one vent producing gas, ash and/or lava. In addition, many of the world's land volcanic vents are present in large numbers in what are called volcanic fields or magma fields with more than one volcano. Like most figures for world geologic processes, the number 500 is an estimate, rather than a concrete number.

Each volcanic vent spews gases into the atmosphere, with steam, carbon dioxide, sulfur dioxide and hydrogen sulfide typically making up the greatest volume of gas so expelled. 2.6 billion tons of carbon dioxide is the minimum estimate released from these 500 active vents and while it is less than the carbon dioxide estimated to be produced by human activities, land volcanoes are only a small part of the story.

Author's note: When figures have been available or easily calculable for a given process, I have used with the most conservative number (that didn't miss key factors such as many do)— the actual amounts of carbon dioxide produced by volcanism may well be 10 to 1000 times larger than the numbers I am listing.

500 land volcanic vents is a large number, but subsurface oceanic volcanoes called seamounts are even more common. A seamount is usually defined as a volcanic mountain on the ocean floor that is at least 3000 feet high. Of the world's 2.5+ million seamounts and literally millions of smaller oceanic mountains, it is estimated that 650,000+ are active during any given year. While this number seems ludicrous, it is likely to be conservative, as the 10% of the ocean floor that has been explored in detail reveals incredible numbers of volcanic vents that range in size from monsters like Krakatau to small vents like Mt. St. Helens. The 650,000+ undersea vents produce an estimated 3.4 trillion tons of carbon dioxide. Again, a conservative estimate.

The Mid-Oceanic Ridge (MOR) is the single largest volcanic feature on the planet. It is a belt of spreading seafloor that girdles the planet much like a softball's stitches and is roughly 47,000 miles long. The Mid-Oceanic Ridge produces an estimated 244 billion tons of carbon dioxide per year. This is also a minimum figure— it is almost certain to be much larger. The actual number may be hundreds to thousands of times that value, but with so little of the rift system explored, one can only be conservative (unlike the GW/CC lobby, who would simply pick the number they want.)

To get an accurate number for the MOR, thousands of data points, each representing a sample taken of the water's dissolved carbon dioxide adjacent the rift at every mile or less, depending upon the level of volcanic activity and number of activity centers in the rift. No one has as yet done that, but given its size, the MOR system may well represent the largest single source of atmospheric carbon dioxide. It would be worth placing sensors along the MOR to do such dissolved gas studies, even if nothing else but to show just how insignificant man's activities really are, when compared to nature.

With the equivalent of billions to quadrillions of natural smokestacks worldwide pumping out incredible amounts of carbon dioxide and other gases, it seems almost silly to credit man for the carbon dioxide present in the atmosphere. Arrogant. Arrogant and silly. What is amazing is that our natural systems *use* these immense quantities of gases, not merely treating them as "pollutants".

While most of the carbon dioxide produced by undersea volcanism is initially dissolved in seawater, a portion of it eventually makes it into the atmosphere. Some of the dissolved carbon dioxide feeds phytoplankton, which are plants living within 100 to 300 feet of the ocean surface in what is called the euphotic zone. (The euphotic zone is defined as the uppermost layer of the ocean— where sunlight reaches.) Some carbon dioxide is precipitated out of seawater as carbonates such as calcite and dolomite or as cement for sedimentary rocks such as sandstone.

3.6 Carbonate Disintegration and its Impact on the Atmosphere

Production of carbon dioxide from the disintegration of carbonate and carbonate-cemented rocks is the proverbial "800 pound gorilla" that is pretty much ignored in carbon dioxide calculations. Release of carbon dioxide from carbonates and carbonate-cemented rock during erosion/disintegration may be the largest source of carbon dioxide in the atmosphere. Approximately 10% to 15% of the Earth's crust is carbonate rocks.

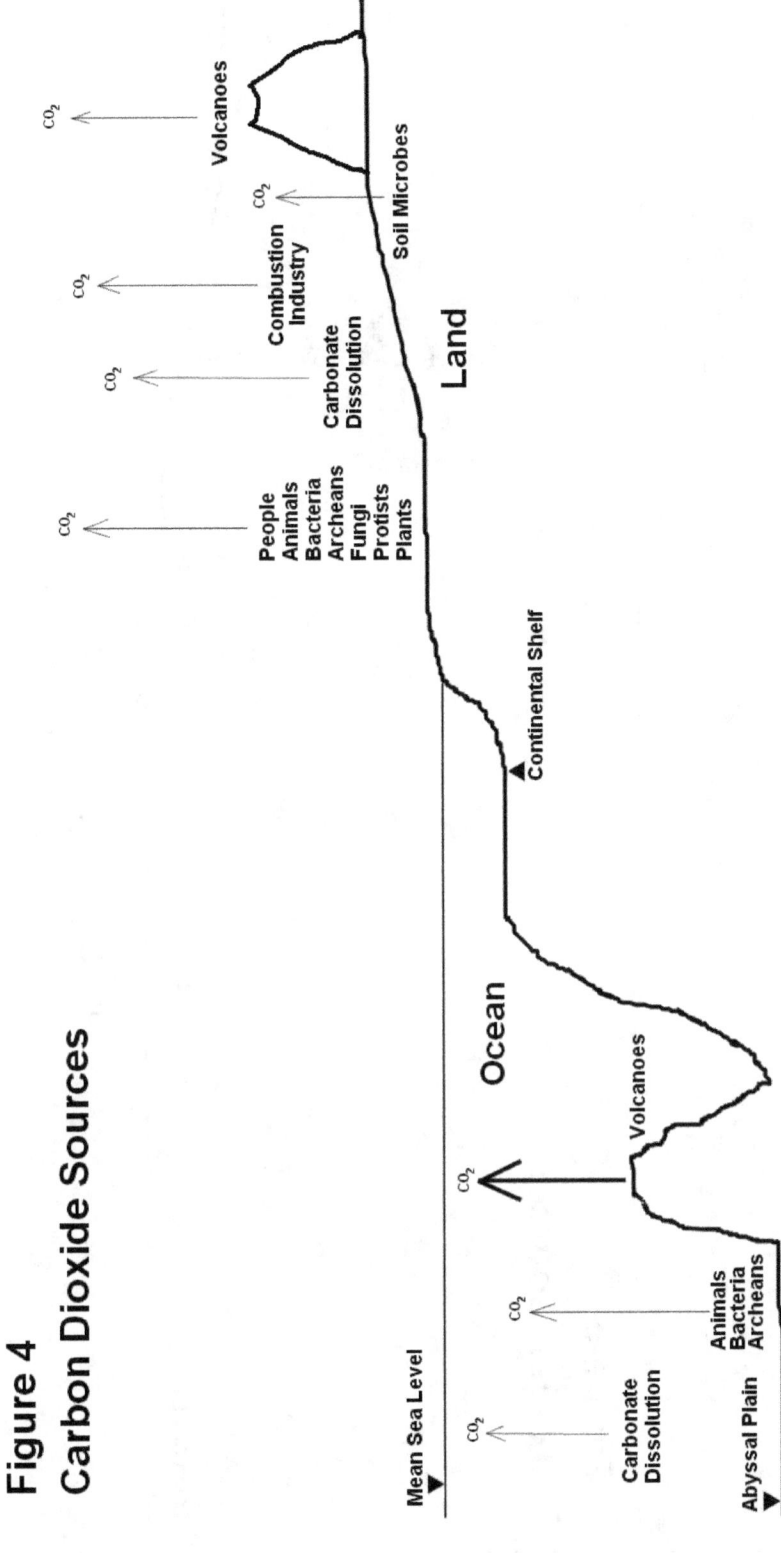

Figure 4
Carbon Dioxide Sources

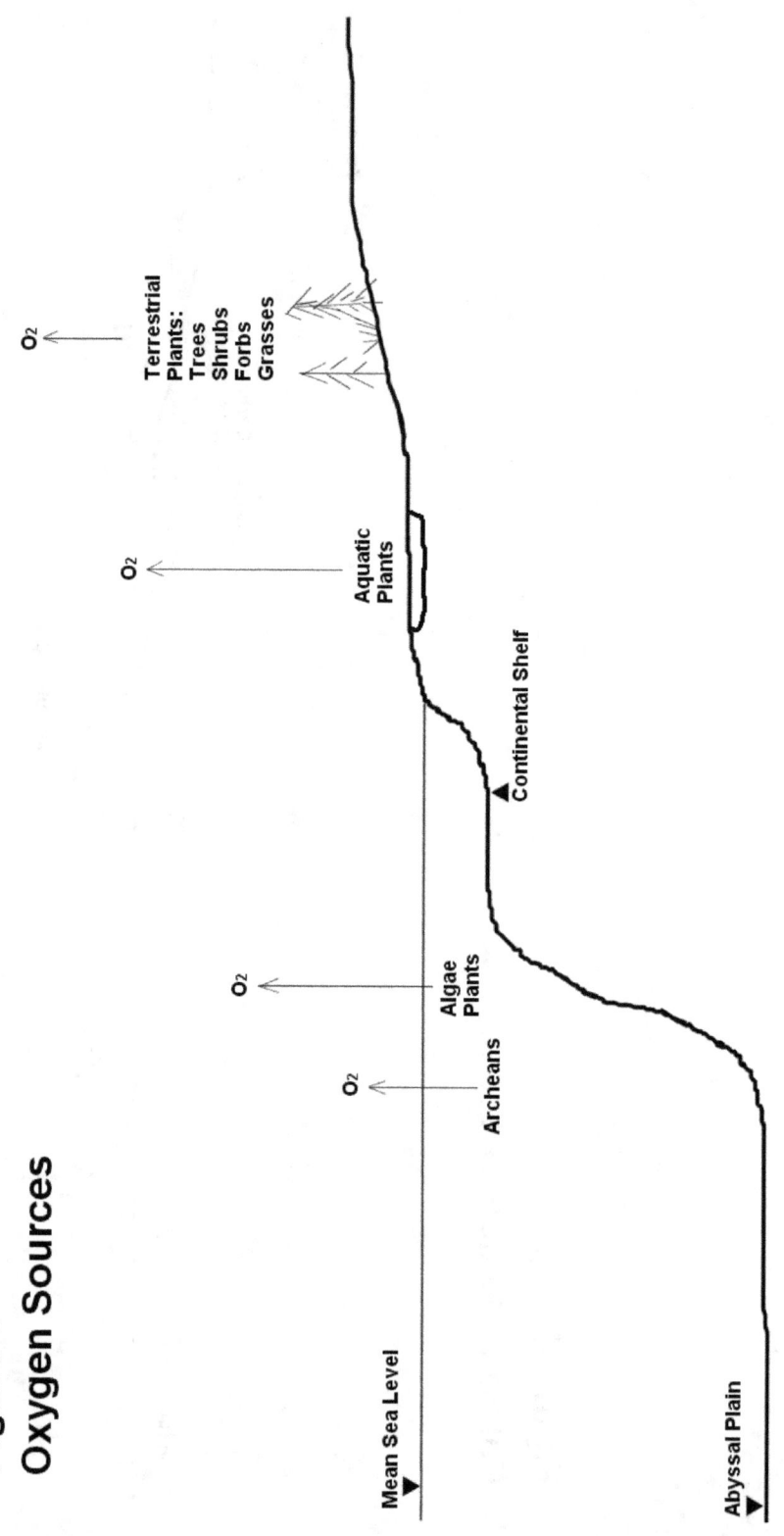

Figure 5
Oxygen Sources

Calcite ($CaCO_3$) is the most common carbonate mineral, often forming rocks composed of pure or nearly pure calcite, such as marble and limestone. Calcite combines with rainwater and carbonic acid (H_2CO_3)— which naturally occurs in rainwater — to form calcium (Ca_2+), carbon dioxide (CO_2-) and bicarbonate (HCO_3-). The carbon dioxide so dissolved generally escapes from the water in short order. The calcium in solution may recombine with bicarbonate or other materials or collect in groundwater or surface water. Many western U.S. lakes and areas of the eastern U.S. have water that is "hard" from dissolved minerals like calcium, magnesium, iron, boron and buffers like bicarbonate.

Other carbonates have similar chemistry. Sodium, iron, zinc, lead, strontium, nickel, magnesium, rhodium and many other elements form carbonates. Carbonates are also present as the cement in approximately 10% of crustal rocks, forming the "binder" holding together sandstones, breccia, conglomerates and other carbonate-cemented sedimentary rocks. (The other common cement of sedimentary rocks is silica). The carbonate cement is disintegrated by the action of carbonic acid just as other carbonate is dissolved.

There are no generally accepted figures for carbonate disintegration-produced carbon dioxide. *Author's note: Isn't that phrase tiresome? I became immensely tired of chasing nonexistent rabbits after months of research.* The rock bottom (pardon the bad pun) minimum figure for carbon dioxide released from carbonate rock and carbonate-cemented rock disintegration on land is 800 billion tons per year. This figure takes into account typical erosion rates for land carbonate rocks around the globe. This **does not** include carbonate rock disintegration in oceanic water as no figures were available — this number only represents carbonate rocks and cement dissolving on continental crust. Including carbonates dissolving in seawater in places where continental crust is submerged or subducting would surely increase the total carbon dioxide produced considerably, but there is not enough data for a calculation.

3.7 Respiration-Produced Carbon Dioxide and the Atmosphere

All living cells produce carbon dioxide as part of the process of respiration. Humans, plants, mammals, insects, reptiles, plankton, crustaceans, fungi, protists, bacteria, soil microbes— every living thing respirates in some form. Some estimates have been made, but the numbers are widely scattered, and as with many other numbers dealing with carbon dioxide and oxygen production and use, there are no generally accepted figures. Given the incredible numbers of living organisms on land, in rivers and lakes, in the oceans and even in inhospitable places like the oceans' volcanic "black smokers" and glaciers, the total amount of carbon dioxide produced by living organisms must be immense. This carbon dioxide eventually makes it into the atmosphere, directly or indirectly, in part or in total, depending upon what organism and where it lives.

3.8 Plant-Produced Carbon Dioxide and the Atmosphere

All plant cells respirate, just like animal cells do, but most do not think of plants producing, rather than merely using, carbon dioxide. There are no figures readily available for plant-produced carbon dioxide and I know of only one study that even takes plant-produced carbon dioxide into account. Given the number of plants on land, in freshwater and in the oceans, the amount of carbon dioxide produced must also be a large number. Respirated carbon dioxide contributes to the total atmospheric carbon dioxide, but it is much less than the carbon dioxide used by plant photosynthesis.

3.9 Human-Produced Carbon Dioxide

This is the area that the Global Warming lobby hammers on. One commonly available estimate from the USGS, EPA and other sources for man-produced carbon dioxide is 29 billion tons annually. However, this number does not appear to include human respiration, decay of sewage or the decay of plant wastes such as farm stubble. In addition, the former East Bloc, China and India have tremendous problems with sewage, garbage and chemical wastes, much of which are eventually broken down by bacteria or other microorganisms. This decomposition releases carbon dioxide, methane, nitrogen and other gases. The actual amount of carbon dioxide produced worldwide by human activities is probably much larger than 29 billion tons per year, but as with many statistics, it's hard to find solid data due to the politics involved.

It's interesting to note that while China has more than twice the industry of the United States with 4 times the population of the U.S. and India has heavy industry comparable with the United States and 3 times the population of the United States, most "estimates" indicate the U.S. as the largest carbon dioxide producer. Again, apparently some sectors of carbon dioxide production, such as those mentioned above, have not been taken into account. If they had, China and India would lead the world in carbon dioxide production figures. When the 2000 coal-electric plants that China is constructing go online in the next five years or less, one wonders if the U.S. will still be considered the "bad guy" when it comes to carbon dioxide production.

Neither China nor India have the enormous area of forested lands that many of the industrial nations possess— *2.5+ million square miles of trees in the U.S. alone.* Russia has more. Canada and some of the eastern European nations have large forested areas. Temperate and boreal forests remove much of the carbon dioxide from the atmosphere, unlike tropical forests which are net oxygen users. If carbon dioxide were the problem that many politicians and environmental groups say it is, they would do well to complain about the nations that produce most of it. Instead, their rhetoric invariably state that the United States is solely to blame and effectively that the U.S. must be the one and only nation to change.

Author's note: when I was working on my masters, a common phrase heard in the news and around college was that the" tropical rainforests were the lungs of the planet." Studies since then have indicated that because of the extreme amount of rot, the rainforests and jungles actually consume slightly more oxygen than they produce. Strange, but true. Again, if carbon dioxide was a problem, then the rainforests would be considered a contributor to climate change. I've yet to see any group go that far. And I hope they do not. The rainforests have many useful plants, animals and other organisms. Even the most ardent Global Warming agitator is not going to fight to eliminate such complex ecosystems like those that exist in South America, Africa, Asia and Australia in the name of "stopping Climate Change". That would be a crime, but then, so is destroying entire economies.

3.10 Photosynthesis and its Effect on the Atmosphere

As most students learn in junior high school biology, photosynthesis is the process by which plants and other photosynthetic organisms take up carbon dioxide, water and sunlight and convert them to sugar and oxygen. This process accounts for most of the oxygen in the atmosphere. No generally accepted figures exist for the oxygen produced and carbon dioxide removed from the atmosphere by photosynthesis worldwide, but the numbers must be enormous. Statistics that I found in various publications varied tremendously and none appeared to take all known photosynthetic organisms to account.

Plants cover most of the Earth's land, from the subarctic Taiga forests of northern Siberia, interior Alaska and southern Chile to the rainforests and jungle of the equatorial regions. Phytoplankton, kelp and other plants fill the uppermost 100 to 300 feet of the oceans, and while statistics also vary widely, most likely account for more oxygen produced (and carbon dioxide used) than land plants.

Some oceanic archeans (organisms similar to bacteria), cyanobacteria (formerly called blue-green algae) and some protists (amoeba-like, water-dwelling creatures) such as the euglena, photosynthesize as well. No figures exist for the amounts of oxygen these organisms create or the carbon dioxide they remove from the atmosphere.

No reference that I have come across lists the effects of crops in comprehensive fashion. Agricultural crops remove carbon dioxide from the atmosphere through photosynthesis just as other plants do, but are completely ignored by the Global Warming lobby. No generally accepted figures exist for crop's conversion of carbon dioxide to oxygen, either. (Although I note that the EPA finally has begun to take crops into account for some of their figures this year.)

Author's note: are you seeing a pattern with so many references of **"no generally accepted figures exist"** *and such? It was frustrating during the research phases of this book that so few of these matters had been studied and numbers calculated for them.* **It says something that so many believe in Global Warming when the basic studies needed have not been done or are so incomplete.**

3.11 Carbon Sequestration and the Atmosphere

Sequestration is a $100 word for the removal of atmospheric carbon dioxide from the atmosphere and the binding it up as carbon compounds in solid form, such as wood or other plant tissue. Carbon dioxide is removed from the atmosphere by photosynthesis and fixed into plant materials. Geologic processes change these plant materials into solid forms such as peat, lignite, bituminous coal, and anthracite. Carbon dioxide is removed to form carbonate rocks, which are buried and make up a small percentage of the Earth's crustal rocks. See Figure 6, Carbon Sequestration.

One of the matters harped upon by the Global Warming lobby is the need to fix carbon by removing carbon dioxide from the atmosphere. There are even businesses now that will "offset" carbon by doing something "green", such as planting trees.

3.12 Carbon Sequestered by Plantlife

Carbon dioxide is converted to plant mass in the form of woody and non-woody tissue. All plant cells contain carbon in their cell walls and internal structures and use carbohydrates for food, derived from photosynthesis. When plants decay or are buried (such as peat forming lignite), the carbon in their tissues is sequestered in the soil as organic matter (such as humus) and in deposits such as lignite. Since coal is the end result of such burying and alteration of plant materials, in a sense, when we burn coal, we are burning long-buried plants.

Figures for plant sequestration are difficult to find and the few figures available are rather suspect. For example, the EPA lists sequestration rates for the United States, but their 2006 claim for the mass of carbon sequestered in the United States was somewhat low— considerably less than the mass of the carbon in the wood put on by the trees of most National Forests *individually*, much less the carbon that other plants turn into carbon retaining plant tissue— shrubs, forbs, grass, etc. According to the USFS, U.S. forests amass from 500 to 2000 board feet of wood per acre per year on average (depending on many factors) and the EPA figure would suggest that the U.S. has no forests at all.

A number of types of green, red and blue-green algae (cyanobacteria) secrete calcium carbonate into their cell walls. (Some apparently use this as protection versus ultraviolet radiation, forming elaborate crystal patterns.) After the algae die, some of their leftover carbonate forms deposits on the bottoms of lakes or the ocean floors. Algae colonies can actually form rock while alive, called stromatolites, which are combinations of trapped bits of sand and rock and the algae's own calcium carbonate.

A bizarre example of this was around 2000 when a Palm Beach landfill leachate collection system was found to be partially blocked by the formation of calcite crystals from algae and possibly bacteria growing in the leachate. (Leachate is the liquid "runoff" from a landfill). Considering how nasty the water percolating through a landfill usually is, it's amazing that anything can survive in it, much less thrive and form rocks from it.

3.13 Carbon Sequestered by Animal Life

Carbon is bound up by the formation of calcite and aragonite (both are CaCO3) in invertebrates forming shells and exoskeletons. Corals, snails, sea urchins, starfish, clams, oysters, crabs, shrimp, prawns, and other animals incorporate calcium carbonate in their shells, which removes carbon dioxide from seawater. Many carbonate rocks are formed from these shells and skeletons. When sea animals die or are eaten, their shells are broken up (or partially digested) and the pieces drift downward through the ocean's waters. Some are compacted into rock and some dissolve into the seawater again or are washed up against the edges of the continents. The white beaches of Florida and the Caribbean are carbonate sands.

Some animals concentrate and excrete calcium carbonate, such as earthworms, but these are only using existing calcite in the soil, not creating it by removing carbon dioxide from the local environment.

3.14 Carbon Sequestered by Carbonates and Bicarbonates

Carbonates are precipitated out as carbonate rock by ocean water and by carbonate-carrying waters elsewhere, forming calcite, dolomite, siderite (iron carbonate) and other carbonates. These carbonates form layers of rock in sheets or masses of mineral with no detectable grain or are composed of larger bodies, ultimately removing carbon dioxide from the atmosphere. An example of the formation of larger bodies of carbonate is the ooids (small, round carbonate bodies that form in thin layers around a central grain) of aragonite that form in the Great Salt Lake in Utah.

Bicarbonates are formed when non-carbonate rocks are dissolved by carbonic acid in rainwater. Carbonic acid (H2CO3) is formed by water mixing with carbon dioxide. The carbonic acid dissolves the rock and forms bicarbonate.

Both carbonate precipitation and bicarbonate formation remove tremendous amounts of carbon dioxide from the atmosphere— apparently far more than plants are able to do. Again, no firm numbers available.

3.15 Summary of Carbon Dioxide Sources

Table 3 gives approximate numbers for carbon dioxide sources, carbon sequestration, oxygen sources, world heat sources and world heat losses. As you can see, there are terrific gaps in the table. This is for two reasons— the information doesn't exist, or the category of information has widely scattered numbers with no number generally accepted by scientists.

Without this information, it's impossible to know the direct effects man has had in the past, is having now or will have on carbon dioxide levels, what effects man has on the Earth's heat budget and the world's climate.

Inversely, given how natural processes completely outclass puny man's efforts, and given what is known of carbon recycling by our natural systems, I submit that it is illogical to assume that man can affect the world climate if he attempts to warm or cool it deliberately, much less accidentally.

Despite all this, the Global Warming/Climate Change lobby merely continues on as if Global Warming and its attendant Climate Change are somehow established facts.

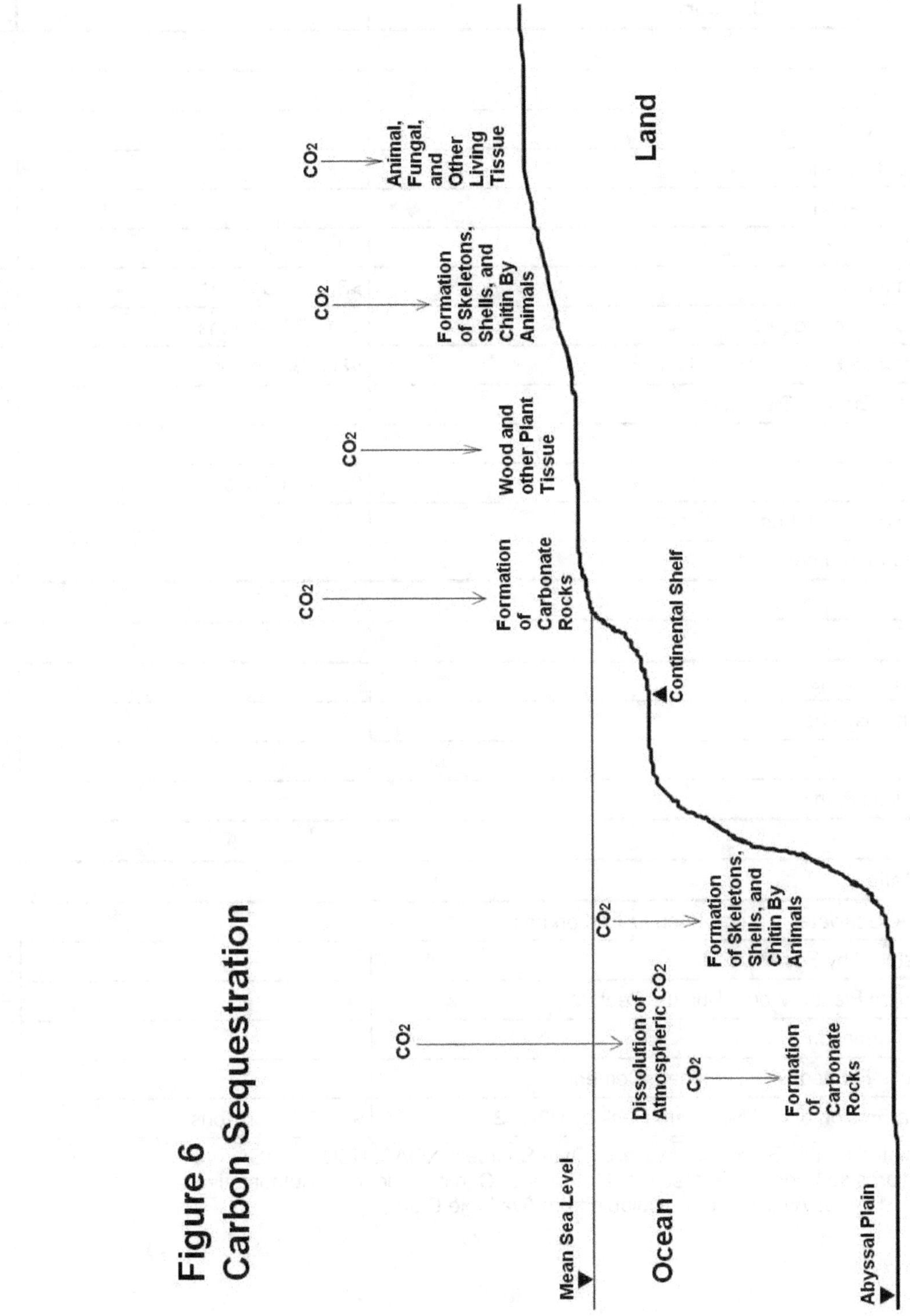

Figure 6
Carbon Sequestration

Table 3 Gas Exchange and Thermodynamics

Category	Estimated Quantities	Notes
CO_2 Sources		
Land Animal Respiration		FWS
Non-Oceanic Aquatic Animals		NFAC
Non-Oceanic Aquatic Plants		NFAC
Phytoplankton Respiration		NFAC
Volcanism- Land	>2.6 Billion Tons	C
Volcanism- Seamount	>3.4 Trillion Tons	C
Volcanism- Mid-Oceanic Ridge	>244 Billion Tons	C
Continental Carbonate Erosion/Dissolution	>800 Billion Tons	C
Oceanic Carbonate Erosion/Dissolution		NFAC
Soil Microbes		NFAC
Human Activities, All	29 Billion Tons	USGS
Land and Aquatic Bacterial, Fungi, Archeans		NFAC
Combustion, Natural Sources		NFAC
Oxygen Sources		
Land Plant Photosynthesis		FWS
Phytoplankton Photosynthesis		NFAC
Oceanic Archeans		NFAC
Non-Oceanic Aquatic Plants		NFAC
Carbon Sequestration		
CO_2 Absorbed into Seawater & precipitated as carbonates		NFAC
Biomass Sequestered by Soil Microbes		NFAC
CO_2 Sequestered by Plants- Wood, Humus, Peat, Muck, etc.		FWS
Carbonate Rock Formation		NFAC
Sedimentary Rocks Formed with Carbonate Cement		NFAC
CO_2 Removed via Erosion of Soil/Non Carb Rock by H2CO3	>12.7 Trillion Tons	C

Notes: C: Calculated for this Book from Available Data Sources- NOAA, USGS, NASA, etc.
FWS: Figures so Widely Scattered that No Number Can be Said to be Authoritative
NFAC: No Figures Available or Calculable from Available Data

Table 3 Gas Exchange and Thermodynamics, Continued

Category	Estimated Quantities	Notes
Heat Sources		
Sun	1362 W/m2 ±5W/m2	NOAA
Total Energy Received by the Earth	348.6 trillion kW per second	
Radioisotopic Heating		NFAC
Compressive Heating of Crust, Mantle, Outer Core of Earth		NFAC
Combustion, Natural Sources*		NFAC
Combustion, Human Sources		NFAC
Heat Losses		
Radiation to Space, Reflected Sunlight and Longwave IR		NFAC
Heat Loss by Atmospheric Gas Losses to Space		NFAC

Notes: NFAC: No Figures Available or Calculable from Available Data
           NOAA: Data from National Oceanic and Atmospheric Administration
           *Forest Fires, Nature-Caused Ignition of Coal, Lignite, Peat, Oil and Gas and Other Combustibles.

4.0 Chapter 4
Other Questions

4.1 What Would Happen if Atmospheric CO_2 Doubled?

Very little, except a small increase in plant growth rates and possibly an increase in carbonate and bicarbonate mineral/rock deposition rates, at least until the atmospheric CO_2 levels dropped and equilibrium was reached once again. The thermal properties of the atmosphere would not be changed by doubling atmospheric CO_2. Even if carbon dioxide increased tenfold, it would not noticeably affect the ability of the atmosphere to absorb, store and release heat. Water vapor accounts for most of the heat the atmosphere can absorb. Nitrogen and oxygen account for the rest. The other gases are in concentrations too small to cause any effect. Carbon dioxide makes up a minuscule amount of the atmospheric gases that doubling it would be like tossing a salt shaker into a marsh.

To change the thermal properties of the atmosphere, it would require such a massive amount of carbon dioxide that it would kill everyone from asphyxiation. The additional carbon dioxide would cause a reduction in the heat retained at night, so it's a matter in question as to whether the reduction in the amount of heat required to raise the temperature due to the carbon dioxide is more or less than the additional heat lost at night. Appendix 3 gives more detail.

Since a carbon dioxide increase will not affect the Earth's thermal regime, it is unlikely in the extreme that Global Warming due to man's activities is even possible.

4.2 What About Other Greenhouse Gases and Their Effects on Climate Change?

What about methane from cows and other "greenhouse gases"? Pretty much every manmade or naturally-released gas is widely advertised to be a greenhouse gas. Nitrogen oxides, hydrocarbon gases like propane, ethane, butane, methane, other gases like ozone, CFCs, and a multitude of others are held to be greenhouse gases. The claim is made that all of these gases increase the amount of heat that the atmosphere retains, which is illogical. All of these gases— except methane— have lower specific heat values than nitrogen, oxygen and water vapor. **None** of these "greenhouse gases" are present in the atmosphere in amounts large enough to impact the atmosphere's thermal characteristics.

Some lump in water vapor as a "greenhouse gas" and if this is so, then water vapor is the only "greenhouse gas" that has the ability to regulate temperatures. Water vapor is unique among the gases in the atmosphere because of the tremendous amount of heat it can absorb and release. If it is a "greenhouse gas" then it is the ultimate "greenhouse gas". Of all of the atmospheric gases, water vapor, nitrogen and oxygen are the gases that absorb 99.99+% of the energy that the atmosphere is able to absorb. Carbon dioxide and other "greenhouse gases'" contributions to the thermal properties of the atmosphere are effectively nothing.

In addition, clouds— despite possibly being made up of a "greenhouse gas" if some pundits have their way — help regulate temperatures by preventing sunlight from reaching the surface during the day and by retaining heat at night.

4.3 What About Ice Cores that Suggest Increased Carbon Dioxide in the Atmosphere?

There is a claim that ice cores from Greenland's glaciers and other areas of "permanent" ice caps, such as the Antarctic ice cap, indicate increasing carbon dioxide. Unfortunately, ice cores in these glaciers cannot be used to determine carbon dioxide levels in the atmosphere of the past. Carbon dioxide and other atmospheric gases sequestered in ice tend to move upward over time, and this mechanism makes ice cores useless for determining atmospheric carbon dioxide. To understand why, one must know how glaciers form and move.

Glaciers are formed by snow falling, building up deep layers that eventually compress into ice. See Figure 7, Glacier Mechanics. Air is trapped by the snow and some becomes incorporated into the ice. With repeated snowfalls, the ice becomes thicker and when it reaches a depth of 150 feet, the ice at the bottom of the glacier becomes plastic, softening and allowing the ice overhead to move. (See Panel 2 of Figure 7.)

The ice at this point begins to flow, very much like a slow-moving river. (See Panel 3.) As the ice flows, the carbon dioxide in the ice moves upward, being squeezed out of the lower layers by the pressure from the ice. (See Panel 4.) Other gases move upward as well, but chemically-inert gases such as argon, carbon dioxide, nitrogen, and helium move the most rapidly— helium escapes very quickly. Chemically-reactive gases such as oxygen move upward, too, but are slower and some may bond to soil and rock particles moved along by the ice rather than escaping to the atmosphere.

This is the same general mechanism that moves much larger gas bubbles upward in flowing lava or a stream. The gas is less dense than the lava or water and moves upward, driven by pressure from below. If you live in an area with recent volcanism such as California, Oregon, Washington, or Hawaii, take a look at a basalt or andesite lava flow that has bubbles. Note that the bubbles are also elongated in the direction that the lava flowed and cooled.

Author's note: I once started a project to determine if Upper Table Rock of the Rogue Valley was an intracanyon basalt flow or was merely eroded into its current shape using this principle. I took soda straws and placed them in the bubbles of exposed lava outcrops, took the resulting direction of the bubble's elongation using a compass and noted the location with a GPS. I ran out of time with only half the project finished, but this was enough to show that the Rock's lava had flowed in a path in a shape around the Rock's horseshoe, as if it had been following an existing canyon like the nearby Rogue River.

Greenland's ice cap is more problematic than the Antarctic. Unfortunately, many scientists are not conversant with Greenland's history. Most of Greenland's ice is of recent origin. Prior to the Little Ice Age, most of the areas where today's core samples are taken, were not covered with ice. The ice that scientists have stated is hundreds of thousands of years old can be no more than a maximum of 650 years in age. Were it not so, farming would have been impossible in Greenland prior to the Little Ice Age.

The upshot of all this is that glacier ice cannot be reasonably used to analyze carbon dioxide abundance in the atmosphere over time as the upper layers of ice will always have more carbon dioxide than the lower layers, regardless of the actual atmospheric levels of carbon dioxide. If carbon dioxide was a massive 10% of the atmosphere 500 years ago and .0033% today, the ice cores would still show the highest carbon dioxide levels near or at the top of the ice. When scientists make claims about the atmospheric carbon dioxide on the basis of ice cores, ignore their claims as the "junk science" that they are.

4.4 The Little Ice Age

The Little Ice Age was a period of time with a much colder climate that lasted from approximately 1350 A.D. to 1850 A.D.. Historians speak of the Thames and New York Harbors freezing during the winters, crop disasters, and the uprooting and movement of peoples due to the changes in the climate. Nations generally didn't move from their homes and way of life unless there was an upheaval or they were forced from their lands. The Turks, Magyars and several other groups moved west from central Asia during this time, likely because the new climate wouldn't support the way they had been living.

There are several theories as to why the Little Ice Age happened, and the most common one is that volcanism increased and poured massive amounts of dust and aerosols into the upper atmosphere that blocked the sun's light. That last appears logical, given what happened after Tambora and Krakatau erupted in 1815 and 1883, respectively. 1816 was described by the people of the time as the "year without a summer" which resulted in crop failures in Europe and North America due to frost. The climate was noticeably cooler in 1884 than 1883, but fortunately, the temperatures didn't drop sufficiently enough to cause major crop losses.

The other common theory is that the sun itself is responsible. That is also logical. During the Little Ice Age there were far fewer sunspots recorded than the last century and there appears to be a direct link between the number of sunspots and the amount of energy the sun emits. Perhaps both sun and increased volcanism played a part. In any case, without more evidence, we cannot know what caused the Little Ice Age. That should give pause to those convinced of Global Warming and Climate Change— if we don't know what caused the drastic cooling of the climate for **500** years, isn't it a tad silly to predict the future climate with such "certainty"?

There have other been recorded periods in the past when the climate turned colder and caused migrations. Some have proposed that a climate change in northern Europe forced the "Sea Peoples" to invade Egypt. I digress, but one migration I've never understood is why the Magogites (later called the Scythians) moved from the wonderful climate of Asia Minor to the Caucasus and later, the Russian steppes. Perhaps the steppes weren't as cold then as today. It's hard to imagine a foe in that time period who would have been able to drive the warlike forebears of the Russians out of Asia Minor.

4.5 Much Warmer Climates of the Past

Given what historians have written about crops and human activities, one has to conclude that the world's climate has been warmer at times in the past. Prior to the onset of the Little Ice Age, Scandinavia was a much warmer place and farming was not restricted to the southern tip of the Scandinavian Peninsula as it is now. Greenland was actually green, and the areas in Greenland where the Norse farmed are currently under ice. The Norse colonies in Canada would not be sustainable today, being extremely marginal for any type of farming and certainly too cold to grow grapes, much less to have wild grapevines, for which Vinland was named. Yet farming ruins have been found in Canada in areas now too cold to farm. Much of Europe was once able to grow crops in locations that are too cold for those crops today.

There are other periods in history that have been suggested as being warmer than now, and it could be argued from a historical viewpoint that the current cold period is anomalous, rather than being unusually warm.

4.6 Don't Problems with Coral Reefs Indicate Oceanic Warming and Sea Level Rising?

No. Most of the Pacific and Indian Oceans' coral reefs are under attack from those living next to them or being starved to death by overfishing by the European Union, Japan and other nations. The pattern usually seen is for the inhabitants of islands to overfish the islands' reefs until no fish remain and then to dynamite the reefs to get the remaining fish and other seafood. The overfishing removes nutrients that support the corals and the dynamiting breaks the reef apart and kills the corals from the shockwaves generated by the explosions. Seawater temperatures don't even figure into it.

Bleaching is the discoloration of coral as it dies from pollution or other attack and is common near the estuaries of industrial nations in or near the tropics, where the rivers carry sewage and chemicals. Florida has bleaching of this kind immediately offshore of some of its river mouths.

Another kind of bleaching occurred in the Caribbean in the early 2000s, when dust carried abroad from dust storms in the Sahara eventually fell back down into the Caribbean, bringing aspergillus with it. Aspergillus is a soil fungus, but upon entering the ocean, it attacked the corals and killed many of them.

Bleaching is claimed to be the result of Global Warming's increasing temperatures, but no reputable study yet has indicated a rise in sea temperatures. A 2002 NOAA study claimed a 0.5 degree increase in seawater temperatures, but 0.5 degrees fell within the instrument error of the instruments available in 1950. There were no comprehensive temperature studies of the oceans back then (or have there ever been, for that matter. Today's certainly aren't comprehensive.). One could argue that a temperature "change" less than the degree of error like that indicates that ocean temps have not changed overall since then, but again, the lack of data renders such claims inconclusive.

The method by which coral reefs build up is fascinating. See Figure 8, Coral Reef Mechanics. The oceanic crustal plates move along, and when a seamount builds up (usually via volcanic eruptions) near the surface or is carried by the plate's movement to close proximity to the surface, corals colonize the seamount. During a coral's life, it creates a calcium-carbonate skeleton and when it dies, other corals build their skeletons on top of it. Over time, the corals build upward, forming an atoll or reef. As the oceanic plate moves, the seamount drops down gradually below sealevel again. As the atoll drops, corals build upward to just under the surface, building the rock up, layer upon layer. Over time, this can result in coral rock hundreds of feet thick. When the reef is killed by overfishing, dynamiting, chemicals, sewage, or other damage, this leaves the island subject to subsidence. Without corals growing upward to fill in the gap between the subsiding seamount and the ocean surface, the island itself appears to sink. Some use the subsidence of islands in the Pacific and Indian Oceans to bolster claims that sealevel has risen, without knowing or worse, knowing but not mentioning how such islands form and what happens when the corals are killed off or damaged.

As often happens, the island nations where this is happening to find it easier to blame the industrial nations than to take responsibility for killing off their own coral reefs.

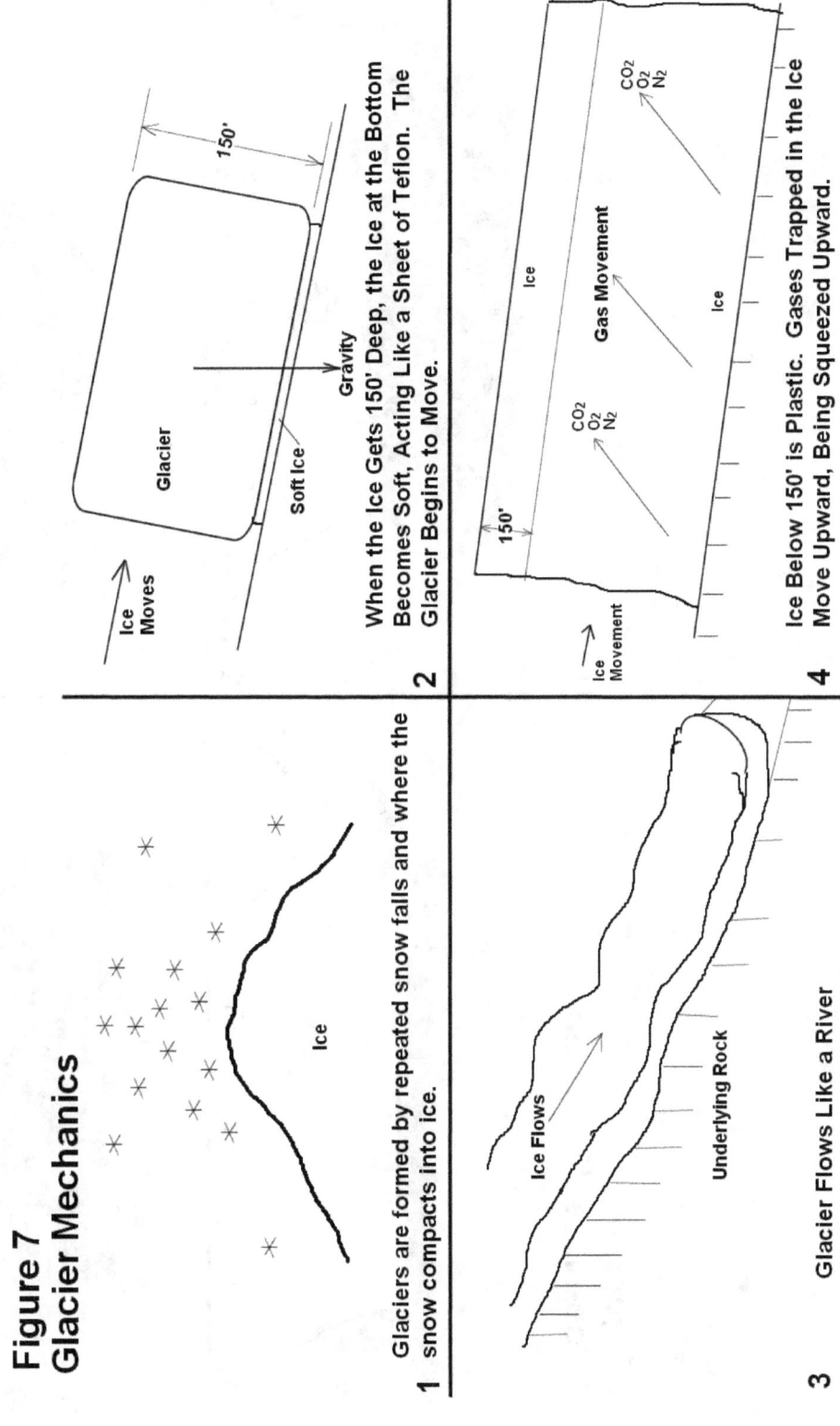

Figure 7
Glacier Mechanics

1   Glaciers are formed by repeated snow falls and where the snow compacts into ice.

2   When the Ice Gets 150' Deep, the Ice at the Bottom Becomes Soft, Acting Like a Sheet of Teflon. The Glacier Begins to Move.

3   Glacier Flows Like a River

4   Ice Below 150' is Plastic. Gases Trapped in the Ice Move Upward, Being Squeezed Upward.

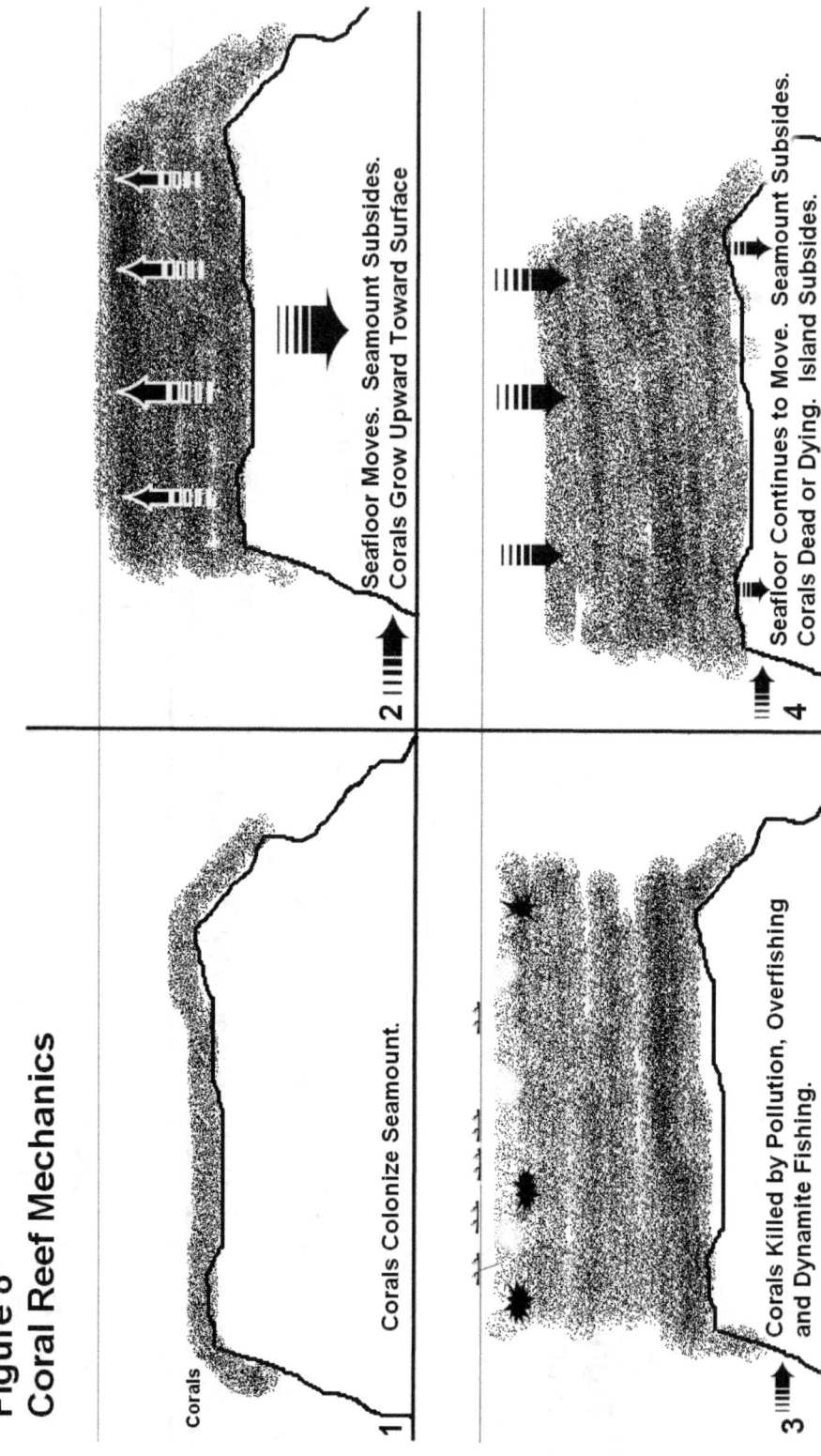

Figure 8
Coral Reef Mechanics

1. Corals Colonize Seamount.

2. Seafloor Moves. Seamount Subsides. Corals Grow Upward Toward Surface

3. Corals Killed by Pollution, Overfishing and Dynamite Fishing.

4. Seafloor Continues to Move. Seamount Subsides. Corals Dead or Dying. Island Subsides.

4.7 Doesn't the "Dead Zone" off Oregon's Coast Arise from Global Warming?

No. The "Dead Zone" (eerie music, please) appears to be the result of high levels of hydrogen sulfide in the water. The media has carefully steered clear of that and blamed Global Warming, both directly and by omission. The "Dead Zone" is frequently brought up by environmental groups and politicians as "proof" of Climate Change.

There are a series of volcanically active areas roughly 150 to 500+ miles west of the southern Oregon coast. The Pacific Crustal Plate and the Juan De Fuca Crustal Plate come together off of the Oregon coast and this has created a fracture zone, with at least three major zones of volcanic activity with a myriad of volcanic vents. Were that not enough, the Juan De Fuca Plate rams up against the Pacific Plate of the Oregon coast. For at least six years and probably much longer, the gases the volcanic vents were spewing forth raised the levels of dissolved hydrogen sulfide in the water off of the southern Oregon coast. At one point, the gases cleared the beaches at Brookings from the awful smell for a day or so. The hydrogen sulfide killed off almost all of the ocean life in the Dead Zone. Then at some point— we have camera images of the result, but don't know when it actually began to happen— hydrogen sulfide-eating bacteria took over and now the area is blanketed with a layer of sulfur excreted by the bacteria. The water has been warmed by heat from volcanic action, but this, too, has been blamed on Global Warming. The television images of the ocean floor showing miles of corpses and skeletons have been used as a rallying cry for Global Warming, but the culprit is <u>volcanism</u>, not Global Warming. The sulfur deposit is the "smoking gun", so to speak.

It is likely that this "Dead Zone" has affected salmonid stocks in the Klamath River of northern California. The numbers of coho, chinook and sturgeon have crashed in recent years, due to a number of causes including stream and oceanic water quality, water temperatures, old, inefficient dams, excessive withdrawals of Klamath and Trinity River waters for agriculture, logging in the 80s, mining and other problems. (Drawing down the Trinity River tributary of the Klamath River by 90% in 2003, resulting in the kill of 30,000+ chinook and steelhead returning to spawn certainly didn't help matters.)

Recently, some have tried to blame the Oregon/California coastal "Dead Zone" on agricultural runoff. This is ridiculous. The "Dead Zone" off the southern Oregon/northern California coast is enormous. At roughly 200 by 300 miles and 2 miles deep, the entire amounts of fertilizer and pesticide used in Oregon and California physically couldn't produce such a large area if they were deliberately spread over it. At worst, such a scenario of dumping hundreds of thousands of tons of pesticide and fertilizer into 60,000 square miles of ocean could produce a short die-off at the surface, but only in the areas where the chemicals were concentrated. If spread evenly over the entire area, they would be too dispersed to cause problems.

The Dead Zone that forms in the Gulf of Mexico, conversely, is man-caused, due to nutrients in the Mississippi and other rivers that discharge into the Gulf. The Missouri-Mississippi-Ohio River System drains 25 states and takes the farm runoff and nutrient loadings to the Gulf. Prior to reforms in water treatment, sewage commonly was also added to the mix and still is added during large storms. To make a complex story short, the excess nutrients in the water cause an explosion of growth in algae and other plants in the water, and when they die off, the resulting rot robs the water of oxygen. This kills off the fish and other animals dependent upon that oxygen in the water.

Author's Note: It is understandable that the uniformed would blame the Oregon/California coastal "Dead Zone" on agricultural runoff. The middle Klamath River has been partially ruined by agricultural runoff and runs olive green even during high flows during the winter. A Secchi disk frequently disappears within 2 feet of the surface or less. It's startling to watch the green water appear to slither over boulders. However, it should be kept in mind that agricultural runoff is only one part of the problem. There are many others and all contribute to both the lack of fish in the river and the terrible water quality. Invariably, the environmental groups want to simplify matters to one or two issues and the quick fixes that they want won't fix the 150 years it required to create "The Big Cruddy", as I call it.

4.8 Hasn't Man Changed the Climate Before?
Local and Regional Climate Changes

Worldwide climates? No. That is beyond man's capabilities. However, there is no doubt that man's actions can change local conditions. Over time, man has changed entire regions from forest or grassland to farming and even desert. The domestic sheep, oddly enough, is the worst offender when it comes to man-caused regional climate changes. If allowed to, sheep will graze every plant down to the soil.

Most of the Mediterranean coast, Fertile Crescent, and Asia Minor region once had **forests**. Think of that: the sand and scrub of Iraq once had *forestlands*. By Hammurabi's time (1790s-1750s B.C.) however, most of the original forest was gone from this vast area and the land was heavily grazed by sheep, preventing reforestation. It's hard to imagine areas like Greece, the North African coast, Lebanon, Iraq, Syria, and the like having large areas of forest. When the forest was removed, the local climate dried somewhat.

To date, Israel is the only nation in the region that has replanted and regrown large areas of forestlands. This has changed the local climate with increased humidity and slightly cooler temperatures. The Palestinians are forever trying to burn the forests down, but thus far, the reforestation of large tracts of Israel has been unique in the region.

A closer example of local climate change is the alteration of the Intermountain West from the 1850s to 1920. The Intermountain Region (the deserts and steppes between the Cascades and Rockies) and northern Great Basin areas of the western United States were described by many as a moonscape by 1920. From the late 1800s into the early 1900s, herds of cattle and sheep were driven through Oregon, Idaho, Washington, and northeastern California, one herd after another.

Author's note: My granddad and great-granddad Page were two who ran sheep and cattle in the Owyhee and Steens areas of southeast Oregon.

Unlike the better-watered Great Plains with its mat-forming grasses that withstand and in some cases, even require grazing pressure from large animals like the buffalo, the Intermountain Region had never been subjected to such intense and constant grazing prior to large-scale grazing in the 1800s. Annual grasses like cheat grass (*Bromus tectorum*) arrived from the Mediterranean and central Asian steppes, taking over what had been grassland inhabited by perennial grasses like Idaho fescue (*Festuca idahoensis*) and native forbs. (Forbs are non-grass, non-woody plants like lupine or sunflowers). While a pain to land managers and land users today, those annuals reduced the soil loss of those times. Many rocky areas with thin soil in central and eastern Oregon for example, would have no topsoil at all today were it not for the humble and hated cheat grass.

This change in vegetation affected the climate locally. Areas with reduced vegetation are hotter during the summers than 150 years ago. The streams downcut, lowering the water tables adjacent the streams, so that the streams' floodplains are no longer watered and thus are now dry and do not support riparian vegetation except immediately adjacent the stream. Conversely, some areas that have "gone to juniper" (the grazing broke up the ground-covering "cryptogamic crust", and combined with fire suppression, which allowed the juniper to take over many areas) are now cooler as the ground is shaded.

The Central Valley of California likewise has had a change of vegetation. As with the Intermountain Region, grazing pressure caused the replacement of native perennial grasses like California Oatgrass (*Danthonia californica*) with cheatgrass, medusahead (*Taeniatherum caput-medusae*), forbs like star thistle (*Centaurea solstitialis*) and nasties of similar ilk that have taken over. Some 20 million acres of California have been taken over by star thistle alone, although recently-introduced insects to control it are making some headway.

Going back to the Near East, Solomon may have been wise in some ways, but he was an early "deforester" (11th-10th Centuries BC). In building Israel's first temple, he had Hiram of Tyre cut endless numbers of cedars, pine, and algumwood (whatever *that* is) from Lebanon. It is recorded that "And the king made silver to be in Jerusalem as stones, and cedars made he to be as the sycamore trees that are in the vale, for abundance." The sycamore-fig tree was very common in Israel, so a LOT of cedars were cut.

Solomon was not the only one— several civilizations cut cedars in Lebanon and eventually, the land was damaged and its climate altered to the point to where only a few small groves of trees survive. The climate is much drier and hotter now in than the forested Lebanon of 3000 years ago. The rivers and streams that used to run from the mountains of Lebanon are mostly gone. It took many generations to reduce the great forests of Lebanon to a few acres, but man did it. Man has also removed most of the great forestlands in China, India, and Western Europe. Only those of Siberia and parts of North America remain in anything even remotely like their natural forms.

The Sahara region of Africa has changed over the millennia. Not so many thousands of years ago, the climate was warmer and the Sahara was much wetter than now. Arab and other African traditions hold that there were rivers and green valleys in what is now the Sahara Desert. This was pooh-poohed by scientists until satellites were able to map what had once been river valleys. Apparently, the climate cooled at some point in the distant past and the rainfall decreased in the region, forming the desert. In the Sahara's case, it is not known if man changed the Sahara's climate. It is remotely possible, given what happened in the Fertile Crescent and the rest of the Mediterranean.

No, there's no doubt that man can alter landscapes and change local conditions. However, changing the global climate is still beyond man's ability.

4.9 What Studies Need to be Done to Determine if Global Warming is Happening? What Data do We Need?

In researching this book, it became painfully clear that most of the studies needed to determine whether Global Warming was occurring as the result of man's actions, simply hadn't been done, and what studies that had been done, had been dramatically oversimplified. Right now, those pushing Global Warming/Climate Change are trying to take a single inch of thread and sew an entire bolt of fabric to it.

Flip back to Table 3 at the end of Chapter 2. Table 3 gives an example of what information is needed for a concrete determination. Note that most of the required fields are not filled with numbers. These matters need to be studied and calculated before one can say with certainty that carbon dioxide has increased or decreased through man's actions or otherwise. Of the numbers that are known or can be calculated, man's activities account for less than 1%. Again, we lack basic data. However, given that man's activities are so small in comparison to other natural processes, it's unlikely that man has affected the global climate.

It would be nice to have another millennium of temperature records for rural areas, but we won't have that for another millennium. Urban areas temperature records are worthless for climate, because there is too much interference from heat absorbed by asphalt and concrete, both in high temperatures during the day and retained heat at night.

A minimum of what is needed:

1. 10,000 fixed buoys spread across the globe taking temperature data for at
least twenty years at the surface, five thousand feet and the bottom.
(This is because if the oceans are actually warming, it would be warming in more than simply the top ten meters in or two places. The oceans are vast, covering 139 million square miles and have an average depth of 2 miles. We have data on much less than 1% of the waters temperatures.) OR

2. 10,000 temperature stations spread across the globe in RURAL areas for at
least one hundred years. Preferably much longer.

The bottom line is that we need real information and we don't have it. A handful of fatally-flawed "studies", incredibly poor science and bad ideology are all that makes up the arguments of the Global Warming/Climate Change lobby.

Why no CO_2 sensor networks? Because the amount of CO_2 required to change the thermal characteristics of the atmosphere enough to cause changes to temperatures would be lethal to many animals and be severely impairing or lethal to humans— several percent. The chronic effects of CO_2 are unknown in humans, with metabolic and adrenal disorders suggested by animal experiments. 10% will kill a person. At the same concentration, plant growth rates take off. I think it's likely that we would notice having trouble breathing, some of our animals dropping dead and our plants growing much faster than normal.

4.10 What Would Happen if the World Warmed?

Given what has happened in the past when the Earth was warmer, world warming would be a boon to mankind. When the world has been warmer in the past, most of the warming took place at high latitudes, dramatically increasing the amount of arable land in Europe and Asia. Imagine being able to farm in northern Siberia— as has been possible in the past at different times. Frost kills far, far more crops than excess heat. Higher temperatures are easily dealt with— many of our staple crops grow better in warm to hot climates.

With more heat, the oceans would warm somewhat. With that comes more evaporation and more rainfall— planet-wide— not just a few areas like the Global Warming crowd claims. Prolonged worldwide drought is actually a sign of cooling.

Increasing humidity tends to limit a given land's temperature extremes, both hot and cold, as well. With warming and more moisture, desert becomes steppe, steppe becomes forest, boreal forest moves into areas that are now taiga and taiga into tundra. Hardwood forest moves northward into areas that are now boreal forests. Imagine the Sahara becoming grassy steppe, as it has been when the world has been warmer. Or imagine Alaska becoming more temperate. With faster conifer growth, like it had around 1000 A.D. or so.

What effect would this have on food supply and land for people to live on? More land to farm and more land to live on.

Much ado has been made about the arctic muskeg and tundra warming up and releasing "centuries" of "pent-up" carbon dioxide. The arctic is less warm than 1000 years ago and so a warm-up is not likely to affect the atmosphere much. In addition, plant and tree species excluded from the arctic from 1350 AD to the present will once again move back in if the climate warms. If nothing else, if the arctic did warm up, photosynthesis rates would increase exponentially, due to the additional plant growth. At the moment, taiga forests grow incredibly slowly due to the very short growing season. That would change with warming. Forest would expand into areas that are now tundra, muskeg might well end up as swamp or forest, and the current, very limited area in the arctic that can support agriculture would expand.

Speaking of the arctic, the winter of 2006-7 saw record ice coverage at both poles and the winter of 2007-8 broke that record. While it may well be a fluke, it doesn't argue in favor for a warming trend. If this trend continues for decades, we'll see a reversal of the trend just outlined, with tundra expanding into areas now of taiga, taiga into boreal forest, etc., as it did during the Little Ice Age.

Returning to the question of what would happen if the Earth warmed: In the U.S., the Great Plains would become semi-arid forest. The Prairies would become hardwood forests (unless fire prevented it, as in many of the world's wetter grasslands). The Appalachians would see southern species move northward, and some species now existing only in the U.S. would be seen in Canada. Canada would see a terrific increase in the area of lands that could be farmed. The current population of Canada, most of which is now in a belt extending to roughly 100 miles north of the U.S. border, could expand northward into areas that now are very marginal for supporting human life.

Not only would the amount of farmland in the subarctic and temperate latitudes increase, but soil fertility would also be increased by higher temperatures. For example, in much of the Central Oregon area, the soil temperature is ten to twenty degrees cooler than soils in the Willamette Valley, some 100 miles to the west. This affects which plants that will grow successfully in the soil of Central Oregon, even when irrigated and fertilized. Forty-five degree soil and temperatures that frequently drop to the mid-thirties at night in August stunt the growth of many garden plants. Composting for a garden is almost impossible in many locations without extreme measures because the climate is so cold and dry. Warmer air temperatures would expedite plant growth. If rainfall doubled to 24 inches, the nearby desert of juniper and sage would become a semi-arid forest of pines and eastside Douglas fir. Water availability would not be the problem that it is currently.

If rainfall increased worldwide, as it would if the world warmed up, imagine the effects upon agriculture in areas where food shortages are common. Increased soil fertility, increased water supplies, and longer seasons for crops. So much for the gloom and doom of the Global Warming lobby. The negative picture of extreme weather they paint is not realistic, given the past warming episodes that the Earth has enjoyed in the past five millennia.

Fresh water is in short supply in many nations and may well become a matter of warfare for nations like Turkey, Syria, Israel, Jordan, and many others. An increase in rainfall would help these nations immensely, both in supplies of water and in allowing more agriculture to take place.

Captain Cook's travels took him to latitude 71° 10' south in January 1774. At this time of year, pack ice normally extends northward to 65 degrees or more! (Surely, the GW lobby doesn't claim we had Global Warming then?!) It was stormy weather and not ice that stopped Cook's journey south. This was more than seventy years before the Little Ice Age ended around 1850. If the world warmed, the Antarctic icepack might retreat completely, as it has in epochs past or it might be confined to the ice shelves of Antarctica. No one knows.

Bizarre possibilities of what might happen in the southernmost latitudes goes back to several maps of Antarctica that predate the modern discovery by hundreds to thousands of years. They show rivers and mountains and the land masses and mountains correspond to current mountain ranges. The jury is still out on whether the maps are real or elaborate fakes. Claims have been made by scientists and historians of both their veracity and their falsity. **If** the maps are real— and that's a big IF— then they would have a profound effect on our understanding of the climate of the ancient world as they indicate Antarctica as ice-free at some point in its history with terrain not unlike other the continents. Given that anthropologists for years claimed that the old world didn't know about the Americas, but Phoenician and Roman coins turned up for years in Indian burial mounds as well as Chinese boat anchors off the west coast, plus Bronze Age copper alloys from the Americas showing up in Europe and elsewhere, I'd wait until the matter is decided before accepting or denouncing the maps. Fact is always better to have than "opinion".

If Antarctica lost its average of 10,000 feet of ice cap, might one think that India and China might be interested in new land for crops to support their millions of people? Other nations would want the mineral riches there. Argentina started the Falklands War to bolster their claims to the Antarctic, despite the limits of current technology to mine Antarctica's mineral and petroleum deposits and the limitations of international treaties. Imagine the land rush if another continent opened up. Germany's desire for "breathing room" in the 1940s would be nothing compared with the desire of China and India—each with more than a billion people— for open land.

4.11 What About Highgrading, Junk Science, and Poor Scientific Reporting?

Highgrading is an old practice wherein data or studies that support a desired conclusion are accepted and data or studies that do not support the desired conclusion are discarded. In principle it's like taking four studies that support the idea that vitamin supplements do no good, but ignoring several thousand studies that indicate that vitamin supplements have some effect upon health. Global warming advocates generally use this technique to "prove" Global Warming is happening.

I define "junk science" as a "scientific" theory or hypothesis that is pushed, either in "studies" or the media, where the "science" itself is directly contradicted by other science. Most Global Warming and Climate Change studies border on junk science or *are* junk science, because the "scientists" conveniently ignore other science— like placing a carbon dioxide sensor on the same island with an active volcano and then stating that the sensor measures *worldwide* carbon dioxide.

Neither materials created by highgrading nor junk science can be considered "good" science—good science can be thought of as that which is supported by the preponderance (majority) of data and studies. Similarly, Global Warming is not good science as it directly contradicts several scientific principles and doesn't fit the entirety of data that we have.

For example, data from temperature stations in the U.S. on average show no change or a slight increase during the time that records have been taken. However, studies of rural temperatures— away from the concrete and asphalt that has warmed city temperatures as cities expanded over the decades— indicate a degree or two less over time. A Global Warming advocate will usually only use temperature records that show an increase, such as those in urban environments that have expanded rapidly over time. This is an example of the highgrading of data.

Science has been held up for decades as being the ultimate problem solver. In some ways it is, but a recurring problem is that scientists have dogmas just as strong as any religion— what is often believed by scientists is the result of insistence that something must be so, rather than the testing of hypotheses. Many scientists are not intellectually honest, deliberately ignoring facts and principles that directly contradict their own ideas and theories.

Global warming, evolution, natural selection and radioisotopic dating are four theories that have lingered on— primarily because many scientists want to believe them— despite volumes of compelling evidence to the contrary. Scientists will not debate these theories because the theories would quickly fall prey to contradicting facts and principles in an open debate. In actual practice, for example, one sees "survival of the luckiest" in nature, rather than "survival of the fittest". (Yes, I know that Darwin didn't say "survival of the fittest", but it is a common enough saying among evolutionists and thus, appropriate to use, being incorrect in its underlying assumptions).

A prime example of junk science is the clinging of scientists to the concept of that the isotopes present in a given rock when it first formed had no daughter products upon formation. Without this assumption, geologic dating doesn't "work". Scientists have shown that one element decays to another, producing elements called "daughter products". Radioactive decay can be demonstrated in the lab. The problem is that scientists take this principle and make the illogical assumption that in any given rock that is analyzed, all of the isotopes that fit the "daughter product" profile were not present in the rock when it was formed. This is contradicted by geochemistry and by the observations of our astronomers/astrophysicists observing supernovae producing mostly radioactive isotopes— theoretically, that is where most elements heavier than helium (and for the most part, lithium) are created and dispersed throughout the universe.

Rocks cannot pick and choose which isotopes of an element they incorporate— all isotopes of a given element are chemically identical. Theoretically, the isotopes created in supernovae eventually fall to Earth and became incorporated into minerals and then into rock. The rocks must have started out with daughter products as they were present to one degree or another when the basic minerals making up the rocks were formed. The scientists want to believe something despite facts to the contrary. As such, the assumption of no daughter products in a rock's initial composition falls into the category of magic or religion, not science, as it can be readily disproven by scientific fact and principle.

Another example: literally hundreds of studies indicate vitamins help in short and long term health (our diets often have minor deficiencies) but several times in recent years, some doctors have frequently leaped on the one or two studies that say they have no effect or even a deleterious effect. That is dishonest.

Global warming is the same way. There are individuals that want to believe it is happening and that it is the result of man's activities— **NO** amount of facts or figures will convince them otherwise. "Don't confuse me with facts, my mind's made up".

To top off the various scientific and junk science debates, there is poor reporting to deal with. When the television runs articles authored by scientists with conflicting information and views, eventually the public begins to tune it out. Eggs are good for you, then they are bad, then they are good for you. All oils are bad, then polyunsaturates are the only good ones, then monounsaturates are the best. And so on. When scientists are proven to be wrong on something that was holy writ by the scientific community, or do a study to "prove" something that everyone already knew— such as "feral cats eat rodents"— science loses face. Telling the public one day that something is healthy and then telling them a week later that it isn't, confuses the public and they eventually treat science with cynicism.

Even worse is when "scientists" tell the public something that is completely wrong. In 2007 I heard an "expert" I'd never heard of before or since on the radio who said that one should have a _total_ sodium intake of less than a quarter gram per day. Not added salt, which might be understandable for a restricted diet, but a _total_ sodium intake for everyone. That would cause severe deficiency in short order. It would be almost impossible to have a diet that met that Forget most vegetables, fruits, breads and meat, minimally processed or otherwise. Even the National Academy of Sciences— which has lost much of its reputation in recent years— has recommended a minimum of a half gram per day for _survival_. Most people have enough wisdom to realize that dropping total sodium intake to almost nothing like that would not be a healthy thing to do.

Many scientists are lousy communicators. If scientists have any interest in getting their message out, they generally don't know how to do it and, indeed, usually don't have the interest in doing it. Most of the public will never read a scientific publication, much less analyze it critically to see if the methodology of a given study is logical or biased in favor of a desired result. To make things worse, the media tends to listen mainly to those scientists who make a big noise, whether what they are saying makes sense or not.

Global Warming/Climate Change is no exception.

4.12 What Was Piltdown Man?

One of the more interesting examples of "science" as the result of insistence rather than testing and observation was Piltdown Man. Piltdown Man was found in 1912 in England and was thought to be representative of an intermediary stage of evolution between primate and man. Apelike jaw with a humanlike cranium. The evolutionists were certain that this was IT. This was the proof they sought. How they crowed! The religions had it wrong and science had triumphed. Nietzsche's story of the madman had been written 30 years early…

Or so they thought. For forty years, few scientists questioned the validity of the find. No one even took a really close look at it, so certain they were what it was. Then, in the early 1950s, a scientist wanted to test the new technique of carbon 14 on the skull. The numbers kept coming up wrong— the bones had different dates. Then he noticed that the bones themselves didn't make sense. A quick study indicated what all the scientists who had looked at the skeleton hadn't noticed— Piltdown Man was made from human and animal bones glued together and painted to look old. And it wasn't even particularly well done.

Didn't anyone dispute Piltdown Man? Yes. Some did. They were ignored or discredited. The scientific establishment _wanted_ to believe that Piltdown Man was real.

Why were they fooled for forty years? Because they saw exactly what they wanted to see in that skeleton and they didn't question what they saw. They saw an image and their minds filled in the rest.

Those who view pornography use exactly the same technique.

Global Warming and it's attendant Climate Change is EXACTLY the same thing. Another Piltdown Man.

Author's note: No one knows who constructed and buried Piltdown Man where it would be found. There are a number of theories ranging from professors, archeologists and anthropologists wanting to make a splash to Piltdown Man being a practical joke. Since Sir Arthur Conan Doyle of Sherlock Holmes fame was at least peripherally involved in the matter, the most entertaining theory was that it was his idea of a joke. Again, we have no real idea who actually perpetrated the hoax.

Section 2 Politics

5.0 Chapter 5 Environmentalism
Who Gains from Global Warming?

So, if Global Warming/Climate Change is not factual, or at the very least, cannot be the result of man's activities, why do people assume it's true and push it— and more importantly, the burden of stopping Global Warming by ceasing all consumption of energy— on others? What do they get out of it?

The politicians pushing Global Warming and its "solutions" hope to gain power and be seen as leaders. There are several politicians in the news that fit this mold. They themselves do not believe in conserving, as evidenced by their profligate use of electricity, gasoline, jet fuel, coal, oils, plastics, and other resources. Despite this, these same "elite" want everyone else to cease using energy and worse, support legislation that would require extreme forms of conservation— by all but themselves. By saying the "right things" about the environment and supporting the "right" legislation, these politicians gain votes. Or at least hope to.

The world media limelight glorifies those who say the "right things", despite the lack of substance to their claims of gloom and doom. Former Vice President Al Gore received the Nobel Prize for "Peace", despite having done nothing for the cause of peace. Indeed, the destabilization of western economies by the currently advocated methods for "stopping" Global Warming and Climate Change would weaken the western nations sufficiently that warlike nations, such as Iran, will be able to act with greater aggression and less fear of retribution for terrorist acts or acts of war. Destroying the western economies will not help anyone.

As such, if there is any lasting effect to Mr. Gore's works, he and politicians like him have done more for creating a climate for war than for preventing it, something apparently lost on the politically-motivated Prize Committee. The emphasis on politics, rather than character or actions is regrettably, not new to the Nobel Prize Committee. Case in point: Yassir Arafat was given the Peace Prize, despite a lifetime of murder that started at age 20 and that culminated in more than 10,000 attacks after Oslo was signed in 1993.

Facts are lost on the world media. They are about producing images rather than substance and right now, Global Warming fits the image they want to portray. Anyone who supports "stopping" Climate Change is "good" and all who do not are "evil". Most politicians want to be seen as "good", whether they actually serve the public good or serve only themselves.

Some politicians in Europe and the Americas hope to use Global Warming as a mechanism to create a world socialist government. Their motive is power, not human "survival" or well being. It is no coincidence that many of the politicians who are serious about "stopping" Global Warming are also socialists of one stripe or another, especially in Europe.

If the idea of man destroying the world via Global Warming is accepted worldwide, then the UN or another international group will be chartered to "regulate" carbon dioxide and to regulate indirectly or directly, all of man's economic activities. This ability to control man's activities represents tremendous power. All industry and commerce from the most industrialized economy to the least technologically advanced societies rely upon carbon-based fuels. From gasoline to firewood. Entire societies can be controlled by controlling their energy supplies. Given the track record of socialism, it's unlikely that such international socialists will be much concerned about human rights, or even basic right and wrong. Certainly the UN has become unconcerned about right and wrong in recent decades.

Think for a moment about that phrase "control man's activities". Isn't that what communist socialists like Stalin or Mao did and national socialists like Hitler did or modern China and Chavez do?

The control of energy inherent in the radical plans of the Global Warming lobby would go very far to satisfying the Sixth, Seventh and Ninth Planks of Marx's Communist Manifesto. The control of transportation, factories and manufacturing, production, agriculture, etc..

A new movement is what environmentalists gain. Environmentalists see Global Warming as a cause in which they and their ideas alone can save the Earth from destruction wrought by Climate Change brought about by the hand of man. Some also see it as a method to bring socialism and world government, ostensibly to bring about "sustainability", restore natural systems and bring about other "natural" goals. By working to "prevent" Global Warming, environmentalists can feel good about themselves and can feel superior to other human beings due to the "rightness" of their cause.

This is not new. In the 1960s, the causes were anti-war and anti-establishment. In the 1980s and 1990s the causes were "save the whales" and anti-nuclear proliferation. Now, we have "Stop Global Warming and Climate Change" as the cause for the 2000s.

There is some irony in that many activists, scientists and politicians blindly accept Global Warming or at least appear to accept it— in order to keep their jobs— while the average person is wise enough to reject the notion. Jokes concerning Global Warming now abound: in offices, stores, groceries, on the street, and at the gas station. When the weather is cold, warm or "normal", people now joke that it must be a symptom of Global Warming. In many ways, Global Warming itself has become a joke. While many accept Global Warming, to most it is a matter of contempt. The jokes one hears almost daily on the street are a symptom of that.

A recent example: winter came a full month early in the Pacific Northwest in 2007— the third earliest on record and the jokes ran along the lines of "must be Global Warming" or "sure be nice if Global Warming hurried up."

Al Gore himself is now the source of many overheard jokes, many of which are not suitable for printing.

5.1 Environmentalism Versus Conservation

What is an environmentalist? What is a conservationist?

For the purposes of this book, the term environmentalist denotes the typical or "radical" environmentalist. "Environmentalist" in this book excludes those conservationists that mistakenly call themselves environmentalists and concentrates on the extreme elements that make up most environmental organizations from the national level to the "grass roots". "Environmental Lobby" refers to the radical environmentalists themselves, their organizations, and those politicians who advocate radical environmentalism.

I've been around environmentalists and conservationists most of my life. Some are good people, some are bad, some are true believers, and some are purely political. The primary difference between a conservationist and an environmentalist is that you will find conservationists actively working to improve forest and stream, while environmentalists primarily are interested in protest, politics, and indirectly, socialism. The term "watermelon" is sometimes used of late to describe many environmentalists, indicating they are socialist— red— on the inside and green on the outside.

An **environmentalist** is a member of the modern environmental movement as a layman, or a member of an organization. An environmentalist is basically an agitator. Environmentalists protest and make a lot of noise about saving nature and "natural" things, but one never sees them cleaning up a stream or planting trees or putting in nets and bales to stop erosion. Most have little practical knowledge of how natural processes work. Since environmentalists are pretty much against everything— any sort of power generation, farming, medicine, logging, industry, commerce, personal freedoms, etc.— and do no work to actually improve natural conditions, it's much easier to be an environmentalist than to be a conservationist.

To be blunt, most environmentalists value ideology over substance and all information must fit their ideology or it's false. "I don't agree with that" is the typical response to factual information, as if it was a legitimate argument. One can counter with "why don't you agree with that? What basis? Please, give me some facts to support your argument", but this will make them think, which can confuse them, frustrate them, make them angry, or much more rarely, make them think more critically of their position. Usually, asking questions makes them angry, because an environmentalist generally has no facts or flawed facts to draw upon and they quickly begin attacking the questioner personally, having nothing but emotion to draw upon. This means that they have lost the argument, but cannot admit it due to false pride.

The environmentalist believes in the preservation of public and private lands in their "natural" state and the changing of lands in non-natural states back into natural states. The obvious problem with that idea is that a given land's natural state is often a matter of presumption, fad and dogma, rather than a true representation of the original state of a land's biology and geology— if the original state of the land is known at all. Western Europe's ecosystems, for example, have been so mangled that there is nothing remotely "natural" left and we can only guess at what the lands were originally like.

There are two general types of environmentalist— those with an education in some sort of natural field (biology, natural resources, wildlife, fisheries, ecology, etc.) and those who are in the environmental movement without any understanding of the natural world. The latter are the most numerous. Very few environmentalists understand natural systems or they would not advocate the positions they do. Environmentalism is at its core a cause to be pursued that gives life meaning. Most environmentalists are into the religions of socialism and Earth-worship to one degree or another. Many add ideas from Hinduism and Buddhism to the mix. Some are vegetarians or vegan. Some branch out into other causes like animal rights. Most smoke marijuana. Those who don't generally have at some point in their lives.

A **conservationist** is one who believes in using natural resources, but takes care of what he/she uses. Conservation is usually not a cause, but is rather a way of life. Conservation has given us most of our practical measures to conserve natural resources, such as recycling. Conservation is independent of religion, but many conservationists are Christian, which fits the Christian ethic of stewardship.

A conservationist is the person that actually does things to improve natural conditions. Improving fish habitat, cleaning up after people litter, removing starthistle and other noxious weeds from infected areas, etc.. These are the people that believe in using natural resources, but also believe in taking care of them. In resource agencies such as the Bureau of Land Management, U.S. Forest Service, National Park Service, etc, they are generally the people hired prior to the Clinton era.

Most, unfortunately, are getting ready to retire, have recently retired or have been involuntarily retired by budget cuts. They saw the Federal government go to heck in the 1990s, beginning with cutting the agencies that actually perform public works by as much as half of their personnel, hamstringing the agencies with additional regulations, and then hiring pretty much only on the basis of skin color and gender without regard to ability or credentials.

As a result, we have a tremendous problem in the land management sectors of the government of incompetence, occasionally ignorance, and worse, ideologies such as socialism, organized racism, contempt for the public and others that are not conducive to public service. There are a few conservationists left in government, but not many. *Author's note: When I worked for NOAA, I stuck out like a sore thumb because I was conservative, a committed Christian, and honestly didn't think much of environmentalists because I'd never seen an environmentalist act constructively. Haven't yet. (Conversely, I have great respect for the conservationists out there. It's getting harder for them to be conservationists, too, what with pressure to be "politically correct" and to abandon scientific principles to say what the environmental lobby wants to hear.) Even worse, I was practical, which occasionally caused me to run afoul of senior management in ways that were downright silly.*

5.2 Environmental Organizations

Most current environmental organizations were originally conservation organizations. The Sierra Club, Audubon Society, National Wildlife Federation and many others were started to protect national treasures such as Glen Canyon and the bald eagle, and to restore our environments that industry, mining, logging, farming or other land uses had polluted or damaged, so that everyone could benefit.

Roughly 30 years ago, several of these conservation organizations began to change course. Having been victorious in advocating conservation laws, they began to advocate preservation, but only preservation for the very few. People no longer were seen as needing forest and wilderness. The wilderness and forest were now considered to be only for the elite and the ordinary person should be excluded. Then, as environmental groups gained power through lawsuit after lawsuit in the 1990s, they abandoned their original mission entirely. Most environmental groups are now purely political. Their purpose is now to stop any use of public lands. Gone is the idea of clean water and air for the common man.

The idea that things must be getting worse in all circumstances is ingrained into the ideology of environmentalism. This, plus the desire for socialist "ideals" and the power that socialism brings to the "elite" few, are two of the most important things to keep in mind when one is dealing with environmental organizations.

For 30 or more years, environmental organizations have been bad-mouthing the United States about perceived environmental problems, but at the same time have ignored the generally much greater environmental sins of other nations. This is laughable— the U.S. was the first nation to begin cleaning up its environment in earnest and started roughly 20 years prior to the next nations to do so— Western Europe. In the 1960s and 1970s, conservation law after conservation law was written by Republicans and signed into law by Democrat and Republican presidents.

Speaking of Europe, it was only in the 1980s that Western Europe finally began to clean up its environment as well. This came on the heels of the realization that half the forests in Europe would be dead within a decade from air pollution, lack of mycorrhizal fungi, soil depletion and other problems if something wasn't done, pronto.

Going back to the U.S.'s groundbreaking conservation laws, the Multiple-Use Sustained-Yield Act of 1960 (MUSY), Wilderness Act of 1964, Clean Air Act of 1970 (CWA), Endangered Species Act of 1973 (ESA), National Forest Management Act of 1976 (NFMA), Clean Water Act of 1977 (CAA), and many others came into being during the 1960s-1970s. The 1960 Multiple-Use Sustained-Yield Act required public forestlands to be used for multiple uses- recreation, timber, husbandry, and others.

Author's note: MUSY was pretty much ignored in the Pacific Northwest from 1976 to 1987 by congressmen and senators who saw dollar signs from rapidly harvesting timber. While often ignored, the original idea of multiple use is a good one. The current mantra by the environmental groups is to prevent any sort of use of the public lands except by the environmentalists themselves, which in effect, would create a personal reserve only for them.

The 1964 Wilderness Act set aside millions of acres of the most remote lands in the United States for everyone to use. The 1970 Clean Air Act eventually made the air breathable again in the eastern U.S..

Author's note: I'll never forget a trip to D.C. in 1983 when the smog was so thick that one couldn't see the sun and visibility was less than 100 feet in any direction. Lung-destroying smog like that is now an extreme rarity and only occurs in places with severely stagnant air like Los Angeles.

These and Acts like them have dramatically improved water, air, and soil quality since the 1960s. For example, modern Air Quality Alerts are only a shadow of the hazard of the Smog Alert smoggy days of 40 years ago, the standards being much stricter for air quality. There is much work left to do, but to constantly complain about the U.S. and compare it to other nations while ignoring those nation's environmental problems is less than honest.

5.3 Blame the United States

The condemnation of the U.S. for "causing" Global Warming is in the same vein. The largest producers of pollution and carbon dioxide— China and India— are ignored, indicating that those involved in such condemnation's real aim is to cause harm to the U.S., both in the world media with constant criticism and by forcing the U.S. to destroy its own economy in order to "stop" global warming.

If the politicians and environmentalists doing the condemning were truly concerned with carbon dioxide and pollution, China and India would be the center of their efforts and propaganda. Instead, they place all blame on the U.S.. The U.S. has been the brunt of nearly constant condemnations by the world media for more than a decade. "Documentaries" quoting "scientific studies" that portray the U.S. as the ultimate villain because of our use of fossil fuels and denigrating our environmental stewardship have been on various world TV and radio networks. Further, various groups and nations have tried to force the U.S. to sign a treaty that NO ONE else to date has honored. The Kyoto Treaty requires a 35% reduction in carbon dioxide emissions by the U.S., while only requiring 5% of the other signatory nations, none of which have reduced their emissions. Notably absent from Kyoto are India and China.

The constant criticism of the U.S. and its citizens by environmentalists falls along the same lines as those "humanitarian" groups that have ignored the problem of muslims killing millions of Sudanese, Nigerians, Chadians, Algerians, Congolese and other sub-Saharan Christian populations since the 1980s. Instead, they complain about the 7 million Israelis supposedly making things difficult for the one billion muslims who want the Israelis driven into the Mediterranean. Or they complain about U.S. prisons where not enough cable TV channels is a major hardship when most of the world's prisons are truly horrific places. No one ever complains about Mexican prisons. Prisoners are regularly beaten to death in Mexico's prisons and jails— many of whom are there for no other crime than not paying a bribe to the arresting officer. The environmentalists do the same sort of dishonest "comparison" by constantly criticizing the U.S. environment and ignoring the horrendous environmental conditions now so prevalent in the third and second world and even in some places yet in central and eastern Europe.

"The U.S. is using too much of the world's resources" is often claimed by the environmental lobby as part of this criticism. Never is any rationale given for that; it's merely an opinion without factual basis. China uses several times as many natural resources as the U.S. does and is increasing this exponentially. Metals, petroleum, wood, coal— China is in the process of building 2000 new coal-electric plants, none of which have pollution controls like those seen in western Europe or the U.S.. Naturally, one never hears complaints from the environmental lobby or the media about China concerning "Global Warning" or resource use. India uses tremendous resources of metals, fuels and food imports and is now held to be the most polluted nation in the world in terms of air quality, food safety, water quality and soil contamination. Again, the environmentalists never mention India.

For third world nations who refuse to correct the problems that are strangling their economies— corruption overriding the rule of law, petty wars, dictatorships, lack of education— resource extraction represents the only hope these nations have to raise their standards of living above mere survival. The environmental groups decry resource extraction and would stop it if they could. They instead seek to deny third world nations the chance to lift themselves up. And they completely ignore the good that they could do by lobbying the governments of those nations to clean up their pollution. Only conservation groups even try to do that thankless job. It's much easier for the environmentalists to demonstrate at home, than do some actual work abroad.

For many environmentalists, the ultimate horror is a world with a high standard of living— why anyone would want to live in places without shacks and dirt floors is beyond them— which is quite peculiar, seeing as they themselves generally come from the more affluent levels of society. But then, most Islamic terrorists are from the upper and middle classes in their societies, too.

Author's note: A couple of years ago, I remember being aghast when an environmentalist stated at a South African conference that it was a such a "good thing" that several thousand children died daily from preventable diseases in Africa. That is truly evil. Those children had names, faces, dreams and had just as much right to live in this world as the environmentalist who made that awful statement has.

5.4 If It Sounds Good, Do It.

How many times have you heard "save the planet", "good for the planet" and somesuch? Environmentalists like to believe that they are "saving the Earth". Even if an action doubles the waste in the landfill by filling it with lead-acid batteries, as long as the right label or phrase is attached to it, it's "good for the planet" or "saving the planet." If it triples the energy cost to manufacture "energy-saving" variants of cars and appliances and produces toxic materials, it's still "Earth-friendly" if it uses less energy on the consumer end. If twice as many people die in accidents in small cars versus large cars, it's "responsible" to drive small cars to save energy. If you have to drive three small cars to move your family around instead of one SUV or minivan, that's "responsible", despite the doubling of the requirements for metals, plastics and fuel. SUVs are "evil".

With that mindset in mind, most words like "sustainability" and "natural" are merely buzzwords, their definitions varying from environmentalist to environmentalist and group to group. "Natural" is seen as anything good in the natural world. The problem with that is that many nasty things are also "natural"— nettles, poison oak, ozone, gamma rays, arsenic, lead and mercury are "natural." Living things killing and eating other living things to survive is "natural." Just because something is "natural" doesn't necessarily mean it's a good thing. I doubt very much that the screaming deer getting its head ripped apart by a cougar would think "natural" was a good thing if it were able to understand the concept of "natural".

"Good for the planet" is in the eye of the beholder. The problem with any action is there is always a tradeoff. For example, when the environmental lobby succeeded in banning offshore drilling off the coast of southern California, the tradeoff was that the naturally occurring oil slicks will continue. The oil often bubbles up off the coast and gums up the beaches. Drilling and removing the oil could stop this, but apparently they'd rather have the oil slicks. They are "natural", you see. Or the tradeoff of preventing the thinning of second-growth forest in the western U.S.. Eventually, this leads to catastrophic fires that replace entire stands, sterilization of the soil and the vaporization of soil nitrogen that took years for the plant roots' bacteria to fix.

The environmental lobby has no understanding of the natural law that every action has a consequence— they have the incredible belief that something can be had for "free". Free energy is bandied about for solar electricity, as if it didn't require several times the materials, chemicals, waste generated in manufacture and total cost than other types of electrical generation. There is nothing free in this world.

For example, the environmental groups have been trying to get away from tree-based paper because other paper is "free" of problems. Paper is made most efficiently with trees, otherwise it would be made with other materials. Weyerhauser experimented with all sorts of plants, shrubs and trees years ago to see what made the best paper. Other companies have done similar studies. Further, using hemp, which the environmentalists invariably want to replace paper, so that they can use it to try to get marijuana legalized, requires a lot of fertilizer, irrigation, weeding and constant tending. Trees require much less maintenance and one can use the tree species that normally grows on the site, which is actually more "natural" than clearing a space of land and growing an alien species like *Cannabis sativa*.

Author's note: don't believe the lie about "industrial hemp" not being used to get high with. Environmentalists have repeatedly claimed that one can't get high with "industrial hemp". That is completely untrue. "Industrial hemp" is the original marijuana plant before people began to selective breed it (and of late, to genetically engineer it) to increase the THC content. The THC content is lower than today's drug-type marijuana, but it's basically the same stuff as the people in the 1960s used to get high with. For those users that don't want to believe that, ask them why people got high from marijuana before it was bred up to its current form.

When environmentalists are not actively protesting, the candlelight vigil is a preferred method to show they care about a particular issue, without actually doing anything to advance it. In truth, I've come to loathe the candlelight vigil, partly because it is usually used to support a bad cause, but on those occasions where it's used for tragedies, it is a way to act like one "cares", without actually putting any effort into helping another person. The candlelight vigil is the epitome of "feel good", as it lacks any substance.

Vegetarianism is another "feel good" fad of some environmentalists. This is despite the increased risk it brings of diabetes, excessive weight gain or loss, the risks of protein deficiency, decreased immune response, heart disease, vitamin B12 deficiency, anemia, and phosphorus deficiency. Vitamin B12 deficiency is characterized by anemia, possible nerve damage, and dementia, according to the Centers for Disease Control. But that is somehow counterbalanced by "feeling" good because they are not eating meat.

Well, if they want to physically feel good while they "feel good", these deficiencies could easily be taken care of with supplements, but many of the vegetarians in Oregon's universities and towns do not take vitamins, protein, or mineral supplements. With predictable results. It's noteworthy that of any nation that keeps statistics, mostly vegetarian India consistently has the highest rate of heart disease. For them, it's a matter of religion, rather than feel-good, but the results are the same.

A perfect example of "if it feels good, do it" is the advocacy of the "mustang" or more accurately, the feral horse. Various groups wanted the "wild" horse to stay on public lands— "aww, they're so cute" or "horses belong on the land"— and got Congress to pass legislation protecting the feral horse. Feral horses cause land managers no end of problems. Horses are not native to the steppes and grasslands west of the Rockies and they do incredible damage. They tear native bunchgrasses completely out of the ground, rather than cutting off bits like the native ungulates. Feral horses trash the streams and watering holes. To maintain the "cuteness" factor, the ones that are ugly or have congenital problems are culled by land managers to maintain the illusion for the public.

Author's note: While walking through some study units in the Salmon Mountains a few years ago, I came across a quarter-acre pond that was a mass of pockmarks from hooves, covered with horse apples and the water, which normally should have been clear, was mud. The grass around the pond was torn up, muddy and dungy. A local feral horse herd had effectively ruined the watering hole for local wildlife— deer, elk, frogs, raccoons, birds, etc..

5.5 Species "Loss"

One of the matters inculcated into biology students is species disappearance. Millions of species are going extinct every year. Or at least that is the claim. Global Warming is now blamed for many of these "extinctions". The problem is that most "species" are merely varieties of species and any minor variation in color, body size, or shape is enough to call something a new species. In the tropics, where a given species may have many, many variations, there is a tendency among biologists to insist that every creature encountered is a different species. (It makes for grants and recognition among their peers and headaches for everyone else.)

In addition, with so many millions of "species" of insects, many are duplicates. It would be fascinating to enter the descriptions of all of those millions of insect "species" into a computer, compare the characteristics and find out how many are the same species, but just having been described by many different scientists and given different names.

Most species are the product not of nature, but of graduate students and professors trying to make a splash, publish papers, and raise money for research grants. For example, there is no difference between the Golden Mantle Ground Squirrel and the Cascade Ground Squirrel. They are identical in activities, eating habits, mating habits, social structure, and physiology. The difference is that one lives north of the Columbia River and the others live south or east of the Columbia. That was "enough" to declare them a different "species".

The same type of "splitting hairs" has been done with western chipmunks. If there is the slightest hint of red in the coat or a nose hair is black instead of white, it's a different species. We have more than a dozen "species" of chipmunks in the western U.S., when at best, all are varieties of the same species. The Eastern Chipmunk hasn't been split and is perfectly adequate for the East half of the U.S.. But watch, soon there will be twenty— one for loblolly pines, one for red spruce, one for white oaks, one for red oaks, one for shrub oaks, one for blue oaks, one for swamp oaks, one for one for the American chestnut (now gone), Alleghany Chipmunk, Lake States Munk, Witch Hazel Munk, Piedmont Chipmunk, Tidewater Chipmunk, Greater Asphaltmunk, Adorable Banditsquirrel, and so on, even though they are the same biologically and move from one area to another. (That's an interesting thought— move a few miles and you change your species... for some animals, that *is* enough to change their species, though.) By physiology, there are four types of gulls on the Pacific Coast. However, eye color, slight differences in the color of feathers and you have a dozen or more. It's nonsense. If the same criteria for species were used for people, we would have **6 BILLION SPECIES** of people.

5.6 A Squirrel's Tail

Here is a hypothetical situation, but it's based upon how things are actually done. A grad student at Humdinger University notices a slight bump in the skull bone of a squirrel. Goes to professor. Professor sees opportunity for a grant. Grant is obtained for doing a study. The study traps and kills all of the chipmunks with the bump plus a thousand other unlucky squirrels. (They could X-ray the critters, but the traditional method for studying squirrel anatomy is to kill and autopsy).

The study reveals that 100 individuals out of the 10,000,000 in Siskyashta County have the bump, now called a "local hemispherical cranial extension". Professor lobbies professional organization. Professional organization declares that it's a new species and these 100 individuals are called the Leastuptownsendsmontegueal.enspugnacious-greatermulticolored Chipmunk, or LUMP-C for short. Given a new genus, too, so it's now called Nogginbumpus stuporii.

An environmental group finds out about the new "species" and sues the U.S. Fish and Wildlife Service because it's not listed as endangered, despite nobody having told the USFWS that the "species" exists. Goes through the courts and finally a judge who knows nothing about how this all came about requires the USFWS to protect it by creating critical habitat.

The area of critical habitat just happens to be in a residential neighborhood. Neighborhood now can't mow lawns, build new houses, use pesticides in gardens or repave the roads. Residents are required to leave out peanuts and other goodies for the chipmunks every day or face lawsuits. Environmental groups sue, blocking every proposed action, reducing the neighborhood to penury. Unable to meet their property taxes due to the legal costs, they try to sell their homes, can't— even to the environmentalists— and gradually begin to abandon the neighborhood.

Two years after this all began, a belated case comes before the local circuit court for "taking" the endangered squirrels because around the time of the listing, some local kids had been shooting them with paintball guns. "Taking" is a catch-all term for any action that is deemed to be bad for an endangered species. Shoot, kill, wound, harass, feed, tease, run over with a car, get attacked by, get bitten by, defend ones'self from, protect one's property from, protect one's cattle from, look at cross-eyed, etc., are all "takings" under the Endangered Species Act. A local biologist notes that the bumps were bruised skull bones resulting from the local kids shooting the squirrels with paintballs and now several kids are now in prison for "taking" the new endangered species.

The environmental groups and scientists eventually realize that chipmunks don't represent a new species, but the environmentalists continue ahead with their lawsuits to protect the species because the neighborhood was especially nice and they see the chance to "protect" it; i.e., keep everyone out except themselves. The scientists don't want to admit that they were wrong and continue to claim the chipmunks are a new species.

The neighborhood eventually is abandoned entirely as no one can do anything without a "taking" and no one will buy property in such a legal mess. By that time, everything is run down and even the environmental groups don't want it, having moved on to other battles. The species goes "extinct" within a year because there are no longer kids around to color them with paintballs. The environmentalists then complain about the loss of the species and alternately blame Global Warming and the Bush Administration for not doing enough to protect the species.

OK, I had a little fun with that, but the way we determine species really *is* that arbitrary. (Anyone in the business of wildlife or fisheries is going to recognize at least some of the references in that story, too.) Public policy is made upon that sort of foundation. There is no set scientific definition for species and because of that, any variation in a species— even if it's just a genetic screw-up in one tiny family group of marmots that produces an extra eyebrow tuft— is enough to call something a species. Species are declared to be species that are not really species and then the free-for-all begins. This is NOT the way to protect ecosystems or to prevent species of animals and plants that are genuine species from going extinct.

Call something a new species and everything is downhill from there. With so many "species", some will die out at the whim of nature and the demise of these "species" is often used as "proof" of Global Warming.

5.7 Alternative Energy and the Environmental Lobby

Unfortunately, nowadays, anything but coal-fired and natural gas-fired electrical generation can be considered "alternative" as all other methods for generating power have been suppressed or ignored. To their credit, alternative energy was advocated by the Carter Administration and Congress in the late 1970s. Tax credits were available for installing solar electric and water heating systems, wind generation, and other alternative energy sources. One by one, though, the tax credits disappeared during the 1980s, primarily because the progressive-dominated congress couldn't stand the thought of any potential source of revenue being lost.

In addition, nuclear power was crushed when the reprocessing of nuclear waste into new fuel rods was stopped by President Jimmy Carter, ostensibly because of terrorist concerns. That such fuel was not in a form easily removed from a reprocessing plant and not easily transportable, eluded the President, which, despite his nuclear engineering degree and time spent in nuclear submarines, is curious, to say the least.

Some states still have credits for home alternative energy, and one might ask why California, which has abundant sunlight, does not have tax credits for alternative energy at home such as solar. As always, talk is more important than action to the environmental lobby.

Hydropower is now considered "bad for the environment" by the environmental lobby because some dams have blocked salmon runs and can change the temperature of a river. The environmental lobby basically advocates removing all dams. The solution to the fish problem is to rebuild or replace the dams to allow for fish passage and for controlling water temperature. Instead, the last major dam system, and indeed, the last major public-works project built in the U.S. is the American River Dam System. The last, despite hydropower producing electricity without pollution and the dams providing flood protection, keeping river transportation corridors open year round, and increasing the abundance of some fish species. Sometimes dams will even open up areas to fish, as is the case with some stretches of the Rogue and Applegate Rivers. Salmon didn't make it farther upriver than Rainey Falls on the Rogue River before dams and hatcheries, for example, as most of the River was not accessible to salmon prior to that.

Wind power hasn't been stopped by the environmental lobby yet, but because the units are noisy and the fan blades kill an occasional bird, it may well happen. All it will take is some endangered or threatened "species" to have a few members fly into a windmill's blades. Wind power was ignored for the most part until the last 4 or 5 years and essentially all wind power projects have been constructed by homeowners, private companies or public utilities.

It should be noted that almost any construction will kill the occasional bird. Why a bird will deliberately fly into an easily avoided and slow moving structure like modern wind generators is subject to debate, but it's enough to turn some environmentalists against it.

5.8 The "Evils" of Logging

The current problem with the firetrap forests of Oregon, Washington, Idaho and northern California has its genesis with the decision by the Carter Administration and the Democrat-controlled Congress of 1976 to abandon sustained yield. As a result, the national forests of the Pacific Northwest were cut as quickly as possible— most of the acreage of public forest lands in Oregon and Washington was cut from 1976 through 1987. Roughly 95% of the non-wilderness, non-roadless Forest Service lands in Oregon (96% of Willamette National Forest, 92% of Winema NF, etc.) and lesser percentages in northern California forests were cut (88% of the Klamath NF) during this period.

Now, 20+ to 30+ years later, most of those cut acreages are firetraps because, while Congress loved the money coming in from timber receipts, they rarely allocated money for maintaining the stands after the initial cut. The environmental lobby has worked tirelessly through lawsuits and demonstrations to stop the cutting or thinning of second growth forest on public lands. As such, they are preventing the rehabilitation of forestlands that they claim to cherish. And putting those forests in harm's way.

While there are many types of forest in the West, the typical situation is that after the initial cutting of the old growth forest, the trees grow back rapidly and very close together. The brush grows up around and under the young trees and depending upon the forest type and local climate, between 15 and 30 years later, one has a stand of small trees crammed together with brush and grass under them, creating a "ladder of fuels." A small fire can quickly grow and engulf the entire stand, trees, shrubs and all. Everything is burned up and the forest has to start over.

The Northwest usually experiences a 2 to 4 month drought during the summer. Everything dries out during this time, creating large areas of fire hazard. The Southwest has occasional droughts as well and generally much higher temperatures. The majority of western forests are now in moderate to severe fire hazard due to heavy fuel loadings. Couple that with dry summer months and high temperatures and all it takes is a spark to set off a stand-replacing conflagration. The nearly 1700 fires set off in northern California the weekend of June 20, 2008 is an extreme example.

The original idea was to cut the stands and then maintain them by thinning them every 10 to 30 years, maintaining spacing between trees and removing the understory. Unfortunately, this has not happened for the most part. Because of cheap imports of old growth Canadian timber, the second growth trees have little or no value and are not logged for that reason. Much ado has been made of Canadian subsidies, but it comes down to economics: it's much cheaper to cut 300 old growth trees than to cut several thousand second growth trees to get the same amount of board feet of lumber.

Much of the environmental lobby has adopted the stance that no logging is allowable for any reason. Most of the fuel reduction projects of recent years have been challenged in court. This is counterproductive as fuel reductions and maintenance of the stands would serve to bring the stands back to late successional status (old, large trees) much faster than letting such huge acreages burn off, requiring the forest to start again from scratch.

Most federal agencies that are tasked with maintaining the forests are starving for money and are not allowed to hire the people they need. The Forest Service, Bureau of Land Management and other agencies cut half to two thirds of their staffs in some states during and since the meat-grinder layoff period of 1992- 2000. At the same time, they have the problem— most of the acres of forest under their control have now grown into dense, "doghair" stands choked with small trees and brush. Firetraps.

Since 2001, Congress has thrown a few bones to land managers for reducing fuels under the National Fire Plan (NFP), but the money has been completely inadequate to treat the millions of acres now in danger of catastrophic fires. Congress absurdly expected the problem to be taken care of in a year or two. Compounding the problem, most of the people who were experienced in fuels management and thinning had been laid off or were retired.

As part of the NFP, the congress did permit the hiring of personnel with National Oceanic and Atmospheric Administration and U.S. Fish and Wildlife Service to work through the backlog of projects, but by 2003, the backlog had been dealt with, and the agencies tasked with maintaining forestlands were not able to produce the studies required for new projects in the numbers that congress expected— not enough people to do the work. The problem was exacerbated by environmental groups suing the agencies on almost every project, no matter what the expected benefits might be. Land-using groups (ranchers, farmers, timber industry, fishermen, Indian tribes, and others) also occasionally sued, slowing matters further. A surprising amount of forest land has been cleared, despite all this, but the majority of Federal forest lands still need maintenance to prevent catastrophic fires and to bring them back to a more natural growth habit.

The environmental lobby could speed the recovery of the Federal forests by using their influence and money to convince Congress to do stand maintenance or even to volunteer to do some of the work themselves. As such, they could be invaluable. Naturally, they will continue to be part of the problem, rather than help solve it.

5.9 Who Profits From Environmentalism?

Most people don't profit from modern environmentalism. Most corporations and government entities have lost money because of environmental fads and trying to appear "green" by adopting foolish policies. The groups that make the most money are the lawyers who litigate environmental lawsuits. The people who lose are pretty much everyone else.

Carbon offsets are one of the few areas of environmentalism that generate cash, but only for a few. These are a joke. The basic idea of a carbon offset is that, to offset energy use, some money is spent on reforestation or rehabilitating lands to convert carbon dioxide to carbon compounds in vegetation. A person who wastes tremendous amounts of energy and generates huge amounts of carbon dioxide can "feel better" about themselves and look good to the environmental lobby by doing this. It's rather like eating only steak and then paying someone to eat one's vegetables.

An example would be millionaire Congressman "X" who is "offsetting" his jet aircraft use of fossil fuels by using a scam corporation that is "environmentally friendly" that does something perceived to reduce carbon dioxide in the atmosphere. In this case, X Forestry, Inc. is planting an acre of trees. Of course, the senator owns the scam corporation, so he's paying himself. The acre of trees would be planted anyway, because the corporation runs tree farms, so it's just a convenient way for a company to generate extra income. Nothing is conserved by carbon offsets, and the atmosphere is certainly not changed.

5.10 The Waste in Global Warming Politics

Money spent on Global Warming and its prevention is considerable, but it's merely a drop in the bucket compared to what we will see if Global Warming is pursued and the western industrial nations are required to forgo the use of energy to "prevent" Global Warming. Literally trillions could be spent in converting economies over to "earth-friendly" technologies, which, because all of the current choices are less efficient than fossil fuels, hydropower or nuclear energy, will result in a drastic reduction of the standard of living.

What could that money be better spent on? Ending diseases such as malaria and tuberculosis— of which each at least a quarter of the world has been exposed to and hundreds of millions are sick with.

Why stop there? Use the incredible amount of wasted money to cure the common cold. End flu epidemics. Cure plant diseases like Dutch elm disease or chestnut blight. Replace coal-fired power plants with breeder reactors that don't pollute the air. Cure AIDS. Eliminate ticks and mosquitoes, worldwide. Restore fisheries in the Columbia, Los Angeles, Smith, Chetco, Klamath, Sacramento, Eel, Mad, Van Duzen and a multitude of other western rivers. Recycle our stockpiles of nuclear waste into fuel. Educate people in the Third World in better ways to live. Send missionaries to our inner cities. Cure asthma, allergies, eczema, lupus and other autoimmune diseases. Convert economies to hydrogen fuel. Clean up Los Angeles. Protect the innocent from criminals. Return 30 million Mexican nationals back to Mexico to force the cleansing of their own nation of its corruption and contempt for law. Find ways to eliminate all human, plant and animal pathogens permanently. Get rid of noxious weeds, get rid of invasive species of plants, fish, mammals, birds, insects, and reptiles. Create viruses to correct genetic diseases in humans. Develop space industries. Develop new solar cells. Develop new energy sources.

The list is endless. Pursuing Global Warming "solutions" will lead to severely-restricted economies with no tangible benefits and waste money that could be better used to help the world's peoples as well as to conserve the natural world.

Author's note: but again, if what Algore and the Watermelons say it is true, that our increase with wildfires is the result of Global Warming, then who am I to dispute it? What are being rated as qualified as an ecologist, biologist, fishery biologist, wildlife biologist, civil engineer, geologist, hydrologist and a number of other specialties to his "vast experience" and "knowledge" of the environment? What are science and engineering degrees, experience working with endangered species (among other things) and knowledge of natural systems versus the ideology, Marxism, lust for power and years of cannabis sativa intoxication of the Global Warming and Environmental lobbies?

Sorry, folks, but when are people going to wake up and notice that our environment, government and public lands are being trashed by zealots both in public office and NGOs, many who can't go a week without lighting up funny cigarettes and think they are saving the world even as they destroy it?!

5.11 What Can You Do?

Take a stand against Global Warming/Climate Change. Call or write your senators and congressmen and tell them how you feel. Call or write your state legislators. Call or write your county, city, and other officials. It works. As a general rule, they assume that for every letter or call, there are at least 100 people who feel the same way who did not call or write.

If you get a form letter, write a complaint about the form letter and send it back with a photocopy of the form letter. (Keep the original). You can generally tell if a letter is a form letter when it doesn't answer the question or concerns that you wrote about. Don't make the mistake of thinking your voice or vote doesn't matter. That is what a lot of politicians want you to think.

A recent example of the public trouncing a bad set of legislation is the 2006 amnesty plan that the Senate tried to ram through. Thousands of people wrote and called to voice their displeasure at yet more incentives for Mexican nationals to come to the U.S.. The Senate didn't pass the bill, stopped by a firestorm of public opinion. Only because people called and wrote letters.

If you don't make your voice heard, it is likely that Congress will begin to pass laws that restrict energy use and supply— such as California and Oregon already have— under the flag of "saving the world from Global Warming." Most such laws restrict the availability of fossil fuels and penalize energy use. Imagine paying $5+ a gallon for gasoline or having gas lines again.

Another great example is the 2008 debacle in Oregon. Oregon already has increased gas costs and use by 30% due to requiring 10% ethanol. This has had the effect of increasing gas taxes by 30% and increasing pollution. How does that work? Simple. When mileage per gallon decreased by 30%, that means that 30% more gallons of fuel have to be bought and burned. More taxes for the state, which is the real reason they did it, despite protestations that it was somehow "good" for the environment. How could a net 20% increase in the amount of the total gasoline used per mile be good? So much for environmentalists thinking things through. They managed to increase the amount of air pollution per vehicle *and* the amount of fossil fuels burned by switching to ethanol. Brilliant! This is only going to get fixed if people complain enough.

Don't make the mistake most voters make of not wanting to do anything. It doesn't take long to write a letter or make a phone call. Be polite, but to the point. If a proposed bill makes you angry, then say/write that it makes you angry. If you won't vote for that official the next time he/she comes up for election because of their legislation or voting for/against legislation that you disapprove of, tell them that. Few things make a politician more concerned than a lot of letters saying they will not be voted for in the next election.

A nation's standard of living is ultimately based upon the use of energy—
all of life's activities require energy, be it for work or play. Transportation systems require energy. Cooking requires energy. Heating one's home requires energy. Mowing the lawn or driving to the grocery store requires energy. Making widgets requires energy. Going to the coast on the weekend or camping requires energy. All of these represent freedoms, both nationally and personally. The ability to get into a car and drive is an incredible freedom that most of the world does not have. Measures to "reduce" global warming by reducing "greenhouse gases" invariably involve the dramatic reduction of the use of energy and would reduce or curtail all of these.

Great strides in energy efficiency have **already** been made by the western nations, especially the U.S.. The U.S. economy has cut its energy use per unit of goods produced by **half** since 1970. See http://www.eia.doe.gov/aer/ep/ep_frame.html Energy use per person has dropped considerably in that time, as well. It's never enough, though, for the Climate Change crowd. The most radical of them will not be satisfied until the world's population is down to 100,000 and all of us live in huts.

Is that what you want? Then get involved.

6.0 Chapter 6
Environmentalism and Socialism

Most environmentalists have socialist leanings, either consciously or unconsciously— many don't even know that many of their buzzwords and phrases have their origins in socialism, such as "social justice". Given what socialism has done to the environment makes its adoption by environmentalists completely senseless.

At least, it's senseless if preserving/improving the environment is *truly* their goal.

The environments of East Germany, Poland, Bulgaria, Romania, China, Vietnam, Cambodia, Laos, and what are now the Czech, Slovak, Tadzhik, Uzbek, Georgian, Armenian, Azerbaijan, Moldavian, Byelorussian, and Russian Republics were severely damaged and in some areas, all but ruined for human use by socialist governments prior to 1991. Given this, why on earth is the modern environmental movement socialist in its ideology?

In 1991, when the Iron Curtain fell, it was clear that Eastern and Central Europe was a mess. Heavy metals from industry polluted East Germany, Poland, Czechoslovakia and Russia. The Poles used to joke that "if you boiled down a ton of cabbage, you ended up with a cannon ball"— reflecting the high lead content of the cabbage, resulting from lead contamination of the soil. Carrots in what are now the Czech and Slovak Republics had to be analyzed for lead before children could eat them. Poland's Silesia had terrible air quality with lead, cadmium, iron, and other metal dusts being released from industry. Emission controls for pollutants were almost unknown before the former East Bloc became independent from the USSR. The terrible air quality led to increased rates of lung cancers, asthma and other preventable diseases.

If the metals weren't enough, sulfur dioxide and sulfur trioxide from bituminous coal burning mixed with moisture in the air to form sulfuric and sulfurous acids, resulting in the acid rain that used to plague Europe and the United States. Coal power was the primary power source of electricity and heat for the entire area of Central and Eastern Europe and there were none of the pollution controls that the west uses. Soils and lakes were damaged by acid rain. Fly ash soot fouled the air.

Chernobyl— "absynthe", in Ukrainian— is one of the worst examples of environmental degradation by socialism. In 1986, the nuclear plant suffered a reactor fire, which poisoned tens of thousands of square miles of land. The plant was unsafe from its initial planning, combining the worst portions of Soviet and Western reactor designs. It was designed primarily to produce plutonium for weapons and power generation was a second thought. It had none of the essential safety features that American and European power plants had and when an test of backup systems was made, one of the reactors caught fire, something also impossible for the non-graphite moderated systems of the West to do. The results we know.

The East Bloc is not an isolated case. Socialism in its many forms has damaged the environment everywhere that it has existed as a government or had influence in government. China, which has almost made the transition from communism to national socialism, has become almost a byword for environmental degradation. Rivers have been wiped out by industrial spills, groundwater made toxic, thousands are poisoned annually from pollution and the air made almost unbreatheable in many cities. Cancers are epidemic in some polluted areas. Coral reefs off of southern China have been damaged by pollution from sewage and industry carelessly dumped into rivers and streams. China was never the poster child for conservation, but the term nightmare now seems appropriate. When the 2000 new coal plants come online, one wonders what the people will breathe.

In Europe and North America, socialism is preventing the cleanup of the environment and the preservation of native species— mammals, amphibians, butterflies, reptiles, and birds and frequently caused the very problems that modern environmental organizations complain of. An American example: at the rescinding of "sustained yield" by Congress and Carter in 1976, the Forest Service was basically ordered to throw the woods wide open to wholesale logging— this was the first act of creating the current disastrous forest situation in western U.S. forests.

Socialists such as Carter and several members of Congress and the Senate eliminated conservation programs such as nuclear waste recycling in 1978, eliminated half of the staffs of conservation agencies in the 1990s, and advocated a huge number of other policies that stopped alternative energy programs almost everywhere in the United States.

It's strikingly hypocritical to pay lip service in advocating for species' and ecosystem survival, while causing the meltdown of such conservationist acts like the Endangered Species Act, National Environmental Policy Act, Clean Water Act, Clean Air Act, and National Forest Management Act— by eliminating the necessary personnel and budgets for enforcing and enacting those laws. But such is the legacy of the "progressives" (socialists) of the 1990s and 2000's.

6.1 Global Fears to bring Global Government

Former Czech President Vaclav Klaus said it well in 2007 when he said that global warming was concealed socialism and stated "Communism has been replaced by the threat of an ambitious environmentalism". Unfortunately, he was right. There are a number of groups around the world that believe that the entire world should be under one roof with a socialist government. One way to accomplish this is to keep the populace in a state of hysteria so they won't notice that their freedoms are being curtailed. Using the fear of a world destroyed by Global Warming to eliminate freedom is one current strategy.

During the 20th Century, any number of ideas/theories were exaggerated or made from whole cloth to stir things up. These in include nuclear winter (1980s), a new ice age (1970s), idiots taking over (the eugenics theory of the 1920s), global drought (1990s and 2000s), pesticides making races of superbugs (1980s, 1990s and 2000s) killing off all food plants, Y2K gutting the world's computers (1998 to 2000), no more trees (1970s to 1990s), half of Europe's trees slated to die by 2000 (1990s), mad cow disease killing everyone on earth (2000s), running out of all resources (1960s to present), running out of food (1960s to present), populating ourselves to death (1900 to present), and a multitude of other disaster scenarios. Sometimes these ideas were blown up in the media by well-meaning zealots, or by a careless media wanting higher ratings, but more often they were spread by people with an ax to grind.

None of the things on that list actually happened as foretold but they all represent the same kind of sensationalism that Global Warming is made of. The Global Warming scare will eventually fade from the scene, and will be replaced with a new "threat" to the world. Some "expert" claiming an invasion from Pluto is imminent because the International Astronomical Union declared them a "dwarf planet", or a 1000km wide asteroid smacking into Times Square, maybe…?

I'm not saying that all environmentalists are bad people. Many are sincere. Unfortunately, the majority are wrong in their science and goals, regardless of sincerity, having a poor understanding of natural systems, or of the true meaning and ultimate results of the socialist dogmas their organizations exegete. Some of them and this is especially true of their leadership, are socialist and want a world government, in order to fix the world's environmental problems. Many grass-roots environmentalists are not aware of that, or of the terrible environmental legacy of the socialist governments of the past.

6.2 What is Socialism?

Socialism has been mentioned several times in this work. To truly understand environmental politics, one has to understand what socialism is.

Socialism is the government type that Karl Marx wrote of in his various works and expounded upon by Engels and other "revolutionaries". The basic idea is that everyone should get the same rewards for their work. A nice thought, but not even once in the history of man, socialist nations notwithstanding, has there been a society in which all were rewarded equally or all needs were met.

Socialism in reality is characterized by restrictions on freedom, inertia, extreme inefficiency, severe bureaucracy, heavy taxation, immorality, lack of individual worth, government control of the economy to some degree, guarantees of various types such as education, transportation, and health care, that are often ignored or reneged upon. Standards of living that vary greatly between the social classes in a socialist nation— despite the socialism's claim of not having social classes— and vary greatly between nations. Warfare—cold or hot— has usually accompanied socialism. Socialism makes great promises to citizens but never has kept them in its entire history.

Socialism is also a religion as well as a government type in that the government takes the place of God. Marx said, "Religion is the opiate of the masses."

To learn more about socialism, see Appendices 1 and 2.

6.3 Bias of the World Media

The global media is the most influential tool the international socialists have. Most television and newspaper outlets in the last 20 years have been brought under the control of the socialists/progressives.

A particularly disturbing example of this is the news that the British Broadcasting Company broadcasts today. Twenty years ago, the BBC was a top notch news organization and frequently provided news on what was going on in the world that our American media wouldn't touch, such as reporting the actual events of the Falklands War. I used to listen to shortwave news radio stations in the late 1980s because the BBC and similar news companies would tell of important matters happening in the world while our American media was enthralled with vapid celebrities. We got Madonna while Vietnam and China had a military border clash with hundreds dead, for example.

Now, however, the news that the BBC produces often has the agenda of promoting socialism and radical environmentalism. Gone is the professionalism and comprehensive news coverage that made the BBC such a good news source in the 1980s. Israel cut ties with the BBC in 2003 because the BBC's bias became intolerable to the Israelis: the BBC effectively ignored the 10,000+ attacks the Palestinians made on Israel after the Oslo Accords were signed in 1993. When the Israelis defended themselves against their attacks or struck back at the Palestinians, the BBC—and many others of the world media— would find a way to shift the blame to Israel, rather than those that had done the initial bombing or rocket attack. The BBC would say that Israel killed X number of Palestinians and, if they mentioned it at all, would place the reason for Israel's counterattack at the very end of the article, almost as if it were not important. The reasons for Israeli counterattacks were generally terrorist incidents such as the bombing of a popular nightclub or a shopping center or Israeli homes and businesses destroyed with bombs or rockets. I finally quit listening to the BBC because of this censoring of the news. Regrettably, most other news organizations have undergone similar transformations from information services to propaganda machines.

Similarly, most articles dealing with the environment have to have some connection to Global Warming thrown in. Early in 2007, I read a copy of Audubon magazine and surprisingly few articles dealt with birdlife, despite the name of the magazine. Most of the articles dealt with Global Warming and Climate Change, rather than birds or matters pertaining to birds. One wonders what John James Audubon would think of the magazine bearing his name.

It is the same with almost the entire American media establishment. During the mid 1990s, Congress voted to rescind the limit on how much of the media a single company could own. Thus, almost all television and radio stations and newspapers are now owned by a few companies. Because of this, most are extremely biased and selective in their reporting of the news. They censor many newsworthy items that they don't like or change the facts of a story so that it fits their worldview. If they are reporting on a candidate or official they like, nothing bad will be reported about them, difficult questions are not asked and bad behavior and skeletons in closets left unmentioned or "spun", which is news shorthand for creating lying.

Sometimes the media ignores large events due to this bias. News organizations are the ultimate censors. In 1992, the March for Jesus in Klamath Falls, Oregon drew 5% of the local population. The newspaper and TV stations didn't write or broadcast anything about it, despite its popularity. The radio stations did carry the story. In 1993, 35,000 people showed up (5,000 of the 35,000 were turned away due to a lack of seating) for the Promise Keepers convention at Portland, Oregon's Civil Stadium. Not one mainstream media outlet even mentioned the convention. Complete media blackout.

A week later, a major television anchor broke the silence and wrote an editorial castigating the Portland media. He stated that any time that Civic Stadium is full, it's newsworthy. In summer 2007, a men's Christian conference was held in Redmond, Oregon and it was completely ignored by the local secular TV and radio stations despite the thousands that showed up. (To their credit, local TV mentioned the 2008 conference).

These three events are examples of mainstream media bias taken to an extreme— where large events with thousands of people are completely unreported because they don't advance the values and agenda of the media.

Author's note: I would not have known about the March for Jesus in Klamath Falls if I hadn't been in it. I wouldn't have known about the Promise Keepers convention in Portland if I hadn't lived there at the time and had a friend who went to it. I would not have known about the Redmond men's conference in 2007 if the local Christian radio stations hadn't carried it in their news segments.

Another example of the media control is to compare the media coverage of Bosnia and Iraq. Most reports concerning the war in Bosnia turned out to be lies. Despite the reports of wholesale genocide, only two situations in which civilians were deliberately killed *en masse* were found after the nation was occupied. The hundreds of thousands killed by genocide never materialized. The war was a disaster from a military standpoint with no objectives completed except Serbian surrender. The media made all of it sound good. In the end, when the Serbians gave up, the truth started coming out, but it is still not generally known by most of the American public, due to the media bias.

Conversely, Saddam killed at least 3 million and we've found hundreds of thousands of his people in mass graves. These were Iraqi citizens who were murdered by Saddam for various reasons, such as being Kurdish, being Shiite, being seen as disloyal, or perceived to threaten Saddam's power. Some were killed simply for Saddam's or his sons' pleasure—shredded, dripped on by acid, dismembered, beheaded and tortured in other ways. The war has been a spectacular success militarily, with a lower casualty rate than any previous campaign of similar type, and a democratic government now exists in Iraq.

The media has made it all sound bad with lies such as "no WMDs", "Saddam wasn't that bad", "Bush lied, people died." Wonder how they explain the 20 tons of WMD weapons of Iraqi manufacture found in Amman, the 100+ acre chemical complex underground in the desert, and the joint nuclear program with Khadafi in Libya? They don't explain it. They simply ignore it.

Environmental reporting is the same. What the media doesn't like, it usually doesn't report. When tens of thousands of scientists from field after field and nation after nation sign a petition stating that man is not to blame for Global Warming, it's ignored completely by the newspapers, radio, and television, in the U.S. and abroad.

Until recently, national media organizations didn't carry a word from scientists who didn't accept Global Warming and Climate Change as fact. Why they have decided to start reporting those who oppose the Global Warming fad is unknown, but it is a welcome trend. It is still quite limited, however, compared to the usual "Global Warming is here" reporting.

In some ways, the reporting of everything as being the result of global warming is diverting us from real problems and is even becoming silly. I'm waiting for a headline like "Bumblebee drops dead in mid-flight! Is it Global Warming?", about the death of a bumblebee, insinuating that it's Global Warming, rather than the result of a sudden velocity loss when it meets a windshield. A real example: we've had a die-off of tens of thousands of honeybee colonies during winter 2007-2008. Most were interbred with the Africanized bee, which cannot withstand temperatures below freezing. For two weeks during winter 2007-2008, there were temperatures in the 20s down into Baja and as far east as parts of Florida. Some media pundits tried to blame it on Global Warming. My speculation is that the real reason was the Africanized bee's inability to survive freezing temperatures. Other factors such as varroa and tracheal mites, cell phone towers and hymenopteran-affecting viruses are not new.

Section 3 Conservation

7.0 Chapter 7
Conservation and Man

Conservation is the use of natural resources, while taking care of them as they are used. Conservation seeks that the resources used will endure for the long term. It's more a way of life than a philosophy or a religion. It's logical to use what we have wisely— not a false wisdom based upon fads and an Earth Centric/Socialism religion like the environmentalists, but upon the realization that we humans live on this planet and must take care of it.

Author's note: Okay, by now you may have decided that I am a "planet-hating", "nature"-hating "industrialist". Nothing could be further from the truth. I've been a lifelong conservationist and have worked both ends of the spectrum— in engineering and as a natural resources management specialist. While working with NOAA to conserve endangered steelhead, coho and chinook in the Klamath River, it was joked that I was doing penance— 6 years before, I'd worked for the Corps of Engineers in testing for projects like the construction of the American River dam system. The engineering experience came in handy when I evaluated a project for impact on endangered/ threatened/sensitive species, as I knew what the construction would do to the landscape and how to mitigate for the effects.

7.1 Stewardship and The Judeo-Christian Tradition

Man has frequently been a lousy steward of the land and sea. This is apparent by the denuding of fish from the oceans, poisoning of the topsoil in Eastern Europe during the bad old days of the Soviet Union, air pollution on every continent, water pollution in many nations and a myriad of other examples.

God gave man dominion over nature, but all too often man has either forgotten to take care of what he uses, or more often, was too involved with bare survival to worry about the long term. God is the ultimate owner of the universe and we've been given control over this small portion of it. Prior to the mid 20th century, most people on the planet were concerned only with staying alive— immediate survival, survival to a successful crop and surviving through the winter. Or making enough at their trade to keep fed. It requires a certain amount of security and spare time to be able to look after the land and the sea. When life is a constant struggle to live another year, it's hard to look beyond it and that is still the lot for many of mankind's billions.

However, the West has been blessed by God with prosperity and we now have the free time and resources to look after the environment. Starting in the 1890s, the formidable modern conservation movement began in the United States. This movement was responsible for such groundbreaking laws as the Organic Act of 1897, Multiple-Use Sustained-Yield Act of 1960, Wilderness Act of 1964, Clean Air Act of 1970, Endangered Species Act of 1973, National Forest Management Act of 1976, Clean Water Act of 1977, and many others.

These have had a tremendous impact on how the average American lives and works. The average air, water, and soil in the U.S. is cleaner than any other industrialized nation and most developing nations. In the 1980s, Western Europe began to clean up their environment as well. The former East Bloc began in the 1990s, but only after socialism lost control.

Responsibility in all things is a recurrent theme throughout the Bible. The idea of land stewardship goes back to the first book of the Bible— Genesis. (All Bible quotes in this book are from the King James Version, or Authorized Version, of 1611).

26 "And God said, Let us make man in our image, after our likeness: and let them have dominion over the fish of the sea, and over the fowl of the air, and over the cattle, and over all the earth, and over every creeping thing that creepeth upon the earth.
27 So God created man in his own image, in the image of God created he him; male and female created he them.
28 And God blessed them, and God said unto them, Be fruitful, and multiply, and replenish the earth, and subdue it: and have dominion over the fish of the sea, and over the fowl of the air, and over every living thing that moveth upon the earth.

29 And God said, Behold, I have given you every herb bearing seed, which is upon the face of all the earth, and every tree, in the which is the fruit of a tree yielding seed; to you it shall be for meat." Gen 1:26-29

Dominion carries with it a lot of responsibility. Yes, man has been given control over the earth, but he has to take care of it. Man's first job was to take care of the garden that God put him in.

15 "The LORD God took the man and put him in the Garden of Eden to till it and keep it." Genesis 2:15

After the Flood, the Lord changed the rules a bit, causing animals to fear man and giving animals to man as food.

1 "And God blessed Noah and his sons, and said to them, "Be fruitful and multiply, and fill the earth.
2 The fear of you and the dread of you shall be upon every beast of the earth, and upon every bird of the air, upon everything that creeps on the ground and all the fish of the sea; into your hand they are delivered.
3 Every moving thing that lives shall be food for you; and as I gave you the green plants, I give you everything." Genesis 9:1-3

Today's farmers, ranchers and herders all have to deal with carrying capacity, which is the number of animals a given area of land can support without degrading the land. One of the first recorded carrying capacity problems is mentioned in Genesis. Abraham and Lot had too much livestock and people for the land to support.

5 "And Lot also, which went with Abram, had flocks, and herds, and tents.
6 And the land was not able to bear them, that they might dwell together: for their substance was great, so that they could not dwell together.
7 And there was a strife between the herdmen of Abram's cattle and the herdmen of Lot's cattle: and the Canaanite and the Perizzite dwelled then in the land.
8 And Abram said unto Lot, Let there be no strife, I pray thee, between me and thee, and between my herdmen and thy herdmen; for we be brethren.
9 Is not the whole land before thee? separate thyself, I pray thee, from me: if thou wilt take the left hand, then I will go to the right; or if thou depart to the right hand, then I will go to the left.
10 And Lot lifted up his eyes, and beheld all the plain of Jordan, that it was well watered every where, before the LORD destroyed Sodom and Gomorrah, even as the garden of the LORD, like the land of Egypt, as thou comest unto Zoar.
11 Then Lot chose him all the plain of Jordan; and Lot journeyed east: and they separated themselves the one from the other.
12 Abram dwelled in the land of Canaan, and Lot dwelled in the cities of the plain, and pitched his tent toward Sodom." Genesis 13:5-12

The Mosaic Law had provisions for allowing the land to rest from growing crops:

3 "Six years thou shalt sow thy field, and six years thou shalt prune thy vineyard, and gather in the fruit thereof;
4 But in the seventh year shall be a sabbath of rest unto the land, a sabbath for the LORD: thou shalt neither sow thy field, nor prune thy vineyard.
5 That which groweth of its own accord of thy harvest thou shalt not reap, neither gather the grapes of thy vine undressed: for it is a year of rest unto the land." Leviticus 25:3-5

Unfortunately, they didn't follow this command. This must have affected their crops. Even the Europeans during the Dark Ages let land lie fallow for a year, so that it could recover some of the nutrients used up by the crops.

The theme of responsibility for the land is throughout the Bible but man's environment is only one aspect of man's responsibility. It should only be a part of one's life, not one's whole life. Use your common sense. So, what can you do at home to conserve energy and resources? Read on:

7.2 Home Conservation

There are a number of things one can do at home to conserve energy and reduce our nation's need for fossil fuels, much of which come from totalitarian nations. Most of these energy conservation tips are easy to implement and they don't require a change in one's basic standard of living. Reducing home or office energy use can decrease the loading on power plants when many power customers do the same, thereby cutting infrastructure costs for electricity and reducing the amounts of coal, gas, water, and nuclear fuel used to generate electricity. Keep in mind, though, that conservation can also be overdone. Don't make it tough for yourself or bankrupt yourself to change the amount of energy you use. Use common sense before making changes.

What *is* common sense? Common sense is wisdom. Think of wisdom as the ability to foresee the results of actions or choices. If I do A, then C, not B will result. If T does M, then N won't happen, because P was prevented. This quality of wisdom appears to be lacking in a lot of people nowadays, seemingly.

Here are some suggestions to reduce the human impact on the environment by reducing water and electrical consumption. Besides conservation, these make sense from a pocketbook standpoint since water and electricity cost the consumer and conserving can reduce the future need for new electrical plants and water treatment.

7.2.1 Energy-Saving Appliances

Refrigerators, ovens, washers, and many other home appliances have energy ratings listed on them. These represent the energy used by the appliance in typical use. Values for kilowatt-hours used per year and the expected energy cost per year are listed. When you look for a new appliance, you can compare them using the energy ratings for different models. Appliances that are particularly efficient have a label called "Energy Star".

Shop around the next time you need a freezer, oven, refrigerator, washer, dryer or dishwasher. First, determine what you need the appliance to do and then look for the most efficient model that fulfills that. Don't buy an appliance just because it's efficient. Don't buy something that is too expensive for your budget, either. Use your common sense. Don't buy an energy-saving appliance unless it makes sense for you to do so.

7.2.2 Fluorescent Lights: Saving Electricity with Better Lighting

Replacing your incandescent light bulbs with fluorescents will cost more initially, but saves money in the long run because fluorescents are more efficient at converting electricity into light than incandescent lights. Fluorescent tubes are filled with a low pressure mix of mercury vapor and a noble gas such as argon or neon. An initial high voltage current causes an arc through the tube. This electricity flowing through the tube causes the gas to give off light, or fluoresce. The light produced contains a large percentage of ultraviolet, which is converted by a coating on the inside of the tube (called the phosphor) into visible light.

I've used fluorescent lamps during and since college, but I've had to be choosy because many of the fluorescent bulbs available produced light that caused eye strain or had an annoying flicker. The inexpensive "sunlight" bulbs/tubes and shoplight bulbs/tubes produce a bluish light that can be unpleasant for some uses. Fluorescent lights have dramatically improved over the years, but one has to choose carefully. Full spectrum tubes/bulbs are still more bluish than sunlight, but are a major improvement over the older bulbs. "Warm white" bulbs/tubes produce a light similar to that of incandescents (although, unfortunately, with more ultraviolet than incandescents) and I've had good results over the years mixing warm white bulbs with full spectrum bulbs in a room.

Fluorescent tubes and bulbs have become much brighter in recent years, but the rating on fluorescent lights was a little unrealistic until 2005 or so. Usually, the manufacturers would multiply the fluorescent tube's/bulb's wattage by five, claiming that a 20W fluorescent bulb was as a bright as a 100W incandescent bulb. Trouble was, if one replaced their lights using that factor as a guide, the result was a much darker room. For most bulbs or tubes, a good rule of thumb is to take the wattage of the bulb and multiply it by three for the equivalent wattage in incandescent bulbs. Even then, the "bluer" tubes tended to have more brightness per watt than the warmer colors of tubes.

The best way to compare brightness is to compare the number of lumens, which is a measure of how much light the bulb/tube produces. If a given model of 23 watt fluorescent bulb/tube is rated at 600 lumens, and a given 100 watt incandescent bulb is 1200 lumens, it will take roughly two of that 23 watt model of fluorescent bulb/tube to produce an equivalent amount of light. In comparing fluorescent tubes and bulbs, you will notice that the amount of light produced by a given bulb/tube varies widely from model to model and sometimes from company to company.

Fluorescent lights still have some improvements to make before they completely replace incandescent lights. Of modern light sources, incandescent lamps such as the humble 100 watt light bulb produce light that is most like the sun, which besides the lower purchase price, is the reason that many still use incandescent. If you live in an area with a lot of voltage changes in the power that your home receives from the electric utility, you will have to replace your fluorescent lights more often.

Author's note: when I lived in Yreka, California, which is just 30 miles south of Ashland, Oregon, we were forever having spikes and drops in the line voltage and occasional outages because the electric utility in Ashland was forever having problems. At that time, most of my fluorescent bulbs lasted less than a year before they'd burn out from the uneven power flow. However, in other places that I've lived, the bulbs/tubes lasted for two to three years before burning out. If you have a lot of problems with the electricity from your utility, you need to factor in the cost of replacing the bulbs/tubes more often before converting to fluorescents.

One disadvantage to fluorescents is that they release ultraviolet light to one degree or another. Some bulbs can actually fade the colors of fabrics over time. There are companies that sell clear sleeves to put over the tubes or fixtures that block the ultraviolet light, while letting the visible light through.

Author's note: when I went to work after graduating college, the first thing I did when I moved into a new place was to change out my regular light bulbs for fluorescents to increase the amount of light in my apartment and lower the electric bill. The overall amount of light in the apartment was increased and I saved money over the long term.

The environmental lobby will eventually notice that fluorescent tubes use mercury vapor and have heavy metals in the phosphor coating. Then fluorescent tubes will be "evil" just like incandescent lamps are now and there will be demonstrations to get rid of them.

7.2.3 LED Lighting

LED stands for Light Emitting Diode. LEDs are solid state, reliable and have low power requirements. They have dramatically improved over the years in the quality and quantity of light produced. Warm white LEDs have been available since the 1990s and some LEDs can produce as much light as a 5W bulb. Fluorescent lights are more efficient at the conversion of electricity into light than LEDs, but this likely will change.

Bulbs using LEDs are available, but they are much more than fluorescents and most LED lights do not produce as much light as an incandescent bulb or fluorescent bulb/tube. The saving grace of LED lights is their low power requirements. Most LEDs have a long life, but as the City of Bend, Oregon, found in 2008, the LEDs in traffic lights may only last a year or two before replacement/repair is needed.

Author's note: Two examples of LED power use and improvements in recent years: A fifty LED lamp that I built in 2003 produces 500 candlepower but only uses five watts of electricity. Better yet, an eight LED, 800 cp. flashlight that I constructed in 2007 uses just 2.5 watts! It's roughly as bright as a 40 watt bulb, but cost me roughly $50 to make. As LED technology improves, I expect LED-using lamps to eventually replace fluorescent lamps and incandescent lamps, but (at the time of this writing) they still need to be developed further in terms of efficiency to compete successfully with fluorescents.

7.2.4 Clothes Washers and Dryers: How to Pick Them

Much ado was made about front-loading washers in 2007, but in comparing them, it appears that many use less electricity per load because their capacity is much smaller than that of most top-loading washers. To wash an equal amount of clothing, some front loaders take **twice** as much electricity and use the same amount of water as a typical top loader because of their small capacity.

Compare models of washers before you buy one. Before you shop, have an idea of how much laundry you need to do per week and then find the model that is the most efficient in the use of water and electricity for what you need the washer to do.

For dryers: again, know how much laundry you need to do and then pick the best one for your needs. As always, use your common sense.

7.3 Insulation and Energy Savings

One of the most important energy and money savers for your home is your insulation. This is true for both colder and hotter climates. While often ignored in the southern U.S., having good insulation makes it less expensive to cool your house or office, just as it helps reduce heating bills in the colder parts of the U.S..

Some homes and buildings can be fitted with better or more insulation more easily than others, so get good advice before changing the insulation or adding insulation. (That would require several chapters to address by itself.) One place to start would be to contact a contractor that specializes in insulation, but make sure he/she is a licensed contractor with a good reputation.

Author's note: There are so many bad licensed contractors out there that I'd want to either know the contractor personally, or know people who have worked with them before I'd hire them to do work for me. There are actually contractors that specialize in fixing the mistakes made by other contractors, so be careful when hiring a contractor. Be careful!

7.4 Recycling

Most cities and towns in the United States have recycling programs, where materials like glass and metal are collected to be melted down and reused. The instructions for recycling are usually straightforward. Some cities require glass, metal, cardboard, and other categories to be separated and others allow everything to be lumped into a "recycle bin" to be separated later.

New York, Maine, Massachusetts, Delaware, Iowa, Maryland, Vermont, Oregon, Connecticut, and Michigan require the recycling of beverage cans and bottles using a deposit of five or ten cents that is refunded when the cans or bottles are returned to the grocery. The "bottle bills" that created these programs worked well as most cans and bottles have been returned in the states that have such programs. It made sense, too, as it was a lot cheaper to melt down glass, plastic, and aluminum and recast it into bottles and cans, rather than make new glass, plastic, and aluminum.

Most other states, like Washington, Florida, Texas, etc., still have most of their cans and bottles thrown into the trash because there are no organized efforts to recycle them. California went halfway, with a "redemption" value, but like most half measures, it doesn't work. Most Californians still toss their bottles and cans into the trash because there are only a handful of places to "redeem" them— few people will drive 50 to 200 miles to redeem their bottles and cans. Some cities have programs to remove the cans and bottles from trash for recycling.

7.5 Water Conservation: the Wise Use of Water

Most of the Eastern U.S. has abundant water, but the western half of the nation doesn't receive as much precipitation. There are a few exceptions, such as Oregon's Willamette Valley, and the northern Pacific Coast, but for most of the West from the Great Plains westward, water is in short supply and conservation is a must.

Author's note: the easiest way to save water at home is to turn off the tap when not in use. This seems basic, but I've been around people at college and other places where individuals turn the water on and let it run down the drain while only using it occasionally.

Low flush toilets or flow reducers in showers and faucets often waste water in actual practice. A faucet that may take only ten seconds to blast off the leftover toothpaste may take a minute or more when fitted with a flow reducer. A low flush toilet may take more than one flush to get the job done which often wastes more water than the older models with larger tanks.

Author's note: I don't know if it still exists, but when I was worked as an engineer, there was actually a minor black market in toilets imported from Europe and Canada, because the low flush models had so many problems.

Taking a shower with a flow reducer installed can add minutes to the shower and may actually waste water because it takes longer to remove the soap due to the reduced pressure of the water. Flow reducers can cause other, more serious problems: if there is a pressure differential between the hot and cold sides of a water system, one can actually back up water from one side to the other. A dorm bathroom at a college had that problem with their bathroom sinks before a student removed the flow reducers the management had thoughtlessly put in. (When one ran both hot and cold, the hot side would push the hot water back into the cold water line because the pressure was so much greater in the hot line, even with the hot on minimum flow.) Backflow of water can cause some serious problems, both to health and equipment by allowing bacteria and funguses to be pushed back into the water line with the lower pressure.

Water can be saved in ways that are more practical than flow reducers and low flush toilets. If you have lawn sprinklers, don't run them more than you need to. For men, when you shave, don't fill the sink— turn on the water when you need to rinse the razor and then shut it off. Cities can use sewage effluent for golf courses and other unimportant uses instead of city water. There are many other simple things to do, such as replacing gaskets when pipes or hoses leak and making sure taps are off when not in use.

In areas with hard water, conditioning can either help or make it more difficult to get things clean, requiring much more water to be used for cleaning and laundering tasks than in areas of soft water. Most water conditioning is still done with zeolites replacing metal ions such as iron, calcium, and magnesium with sodium and while this keeps the scale from the pipes, in some cases it makes it harder to remove soap from clothing and hair than the original hard water.

Depending upon the composition and hardness of the water, a system that removes the positive charge of the metal ions can work better. Carbon filters work well, but can require frequent replacement if the water is especially hard. Reverse osmosis filters are more expensive than many conditioning systems, but they remove all of the metal ions and anions, saving the pipes by preventing scale and making the water taste better. Steam distillation is the most expensive treatment for hard water, and generally not practical except in small amounts for use in drinking or as an ingredient in food.

One bit of lunacy in our business world is the recent insistence of several motel chains (such as Motel 6 in Oregon and northern California) using water softeners in places where the water is already soft. The result is taking clean, soft water and making it saline by adding salt, which makes it harder for guests to get clean, adds to the motel's costs in soap and detergent and leaves residues in their linens and other washed materials. Not to mention a given motel has to buy a lot of zeolites to process the large amount of water that it uses and has to increase the amount of water that it uses to offset the problems of the hardened water. The motel ends up increasing the amount of sewage wastewater it produces as well as making that wastewater more saline and increasing the amounts of phosphorus and other chemicals as a result of the greater amounts of cleaning chemicals needed. The local sewage plant has to process the extra wastewater and chemicals and this gets passed on to the ratepayers.

The problem with domestic water-saving measures— at least in the western half of the United States— is that they save only a little water. The water-saving ball is mainly in the agricultural users' court as agricultural uses literally use more water than all other uses combined. For example, in California, 85% of all water used in the state is by agriculture. At least a third of this could be saved through the piping of ditches and canals and the elimination of wasteful irrigation practices such as flood irrigation.

Author's note: A farmer friend in the Rogue Valley was justifiably proud that he saved 50% of his water use when he changed from flood irrigation to sprinklers a decade ago. If more of our farmers were water-wise, then many of our water problems would not exist today.

8.0 Chapter 8
Alternative Vehicle Fuels

Several alternatives to gasoline and diesel fuel are available and could potentially replace them as vehicle fuels. All of these alternative fuels have disadvantages, however, just as the fuels they replace have downsides. Burning number two diesel produces smog, and both diesel and gasoline have prices that are largely independent of supply and demand.

Author's note: the environmental lobby, which has occasionally advocated alternative fuels, seems to have forgotten that all of these except hydrogen create carbon dioxide when burned. So, any alternative fuel that they currently advocate that produces carbon dioxide will eventually be condemned by them.

Biodiesel, ethanol, methanol, coal diesel, propane, natural gas, and hydrogen are alternatives to some of the current petroleum-based fossil fuels. Gasoline and kerosene can be made from garbage, which has become a viable alternative, as well. For the moment, we will call gas and kerosene substitute fuels made from waste biogasoline and biokerosene.

The performance of alternative fuels in vehicles varies, but a rule of thumb for easy comparison is that hydrogen and natural gas will yield two thirds of the mileage of gasoline, ethanol one third, methanol one quarter, and propane yields roughly the same mileage as gasoline.

Propane and natural gas burn clean, producing only water and carbon dioxide exhaust. This renders a vehicle's smog control equipment useless— these devices reduce mileage without decreasing pollutants in a vehicle burning either fuel. However, given the often arbitrary and even counterproductive regulations of our state governments, some states may well require smog control regardless of a vehicle's fuel. If you plan to change your vehicle's fuel, check your state's regulations prior to making any vehicle modifications.

8.1 Biogasoline and Biokerosene

Changing World Technologies, Inc. has developed a method for converting almost any carbon-based waste product, from plastics to paper to sewage, into an oil which contains mostly gasoline and kerosene, which is easily separated into those fuels. The company has constructed a plant in Carthage, Missouri, to turn Con-Agra's turkey feathers and guts into oil.

It has worked very well and the oil produced is significantly lower in cost than oil purchased on the international market. There have been problems with the plant releasing foul odors, but this can be solved relatively easily. The genius of the process is that it can take materials that we don't want— carbon-based garbage of most kinds, including paper, spoiled food, inedible parts of food plants and animals, sewage, wood, bark, agricultural waste, forest industry waste, plastics, and chemical wastes and turn them into useful oils. Every metropolitan area produces a "stream" of waste that is tossed into landfills or is processed through sewage plants and these could be turned into fuel through this type of process.

Case in point: the City of Portland, Oregon, sends its solid waste to the small town of Arlington, some **135 miles** away, where it is placed in landfills. As the metals have been removed, most of the garbage could be converted into gasoline and kerosene to burn in cars and jets. However, this is unlikely to happen as the state committed to reducing all fossil fuel use in 2007, despite having actively prevented most alternative energy sources.

Author's note: it was my misfortune to live in Portland for a couple of years during the mid 1990s, and their recycling program then was schizophrenic at best, the rules changing every month or two. It drove a lot of people nutty who would ordinarily have been gung-ho about recycling. No one could plan ahead or establish a routine because the rules were constantly changing.

All of the solid wastes from the cities in the Willamette Valley, the Rogue Valley, the Umpqua Valley and Central Oregon could be routed to Arlington via rail or by highway, where the waste would be transformed into gasoline and kerosene. Estimates vary, but a project of this type could provide a third to half of the gasoline and kerosene needs of the state at a fraction of the current cost of gasoline and kerosene. Such a project could also be a money-maker for the state of Oregon, reduce the prices for fuels and solve our waste disposal problems. However, if such a project was initiated, it would be necessary to prevent the existing oil companies from purchasing or controlling the plant, as the price per gallon would be increased to match that of fuels refined from Alaskan oil.

Due to price fixing, the Pacific Northwest usually has the second highest prices for gasoline in the nation— second only to California. It would be in Oregon's interest to pay for an oil from garbage plant at Arlington. It is symptomatic of fuel prices being independent from demand that Oregon's gasoline prices are so high, despite the fact that the oil the Pacific Northwest uses originates in Alaska, which is much closer than the Middle East oilfields that feed most of the nation's refineries— the oil pulled from the ground in Alaska is considerably cheaper per barrel to provide to the consumer than the exorbitant cost of oil purchased from the Saudis. Boise, Idaho, gets its fuel from refineries in Utah and if the prices were actually based upon supply and demand, would have much cheaper fuel than most of the nation due to the relatively low demand and the short distance from refinery to market.

In 2008, the price of oil went to $140 per barrel and remained there for a time. Gas went to $4 a gallon. However, when oil dropped to $100, after two months, the price for gasoline was still $3.75, rather than dropping to its previous level of $3 or so at $100 per barrel. It should be noted that, despite the doubling and then tripling of price, according to the DOE, demand was effectively static in 2006, 2007 and again in 2008. Consumers reduced their use in 2005, when the prices first began to skyrocket, consolidating their trips. This reinforces the contention that the price of gasoline and diesel have no relation to supply and demand. In the case of a product that all must have— petroleum to run the cars and trucks— one is at the mercy of those producing the product. Thus, producing gasoline in Oregon makes sense as it introduces competition in a market that has none. Especially when it uses a waste product that otherwise would cost money to dispose of.

Author's note: I've suggested this trash to oil project to multiple Oregon legislators. None were interested. This in not surprising— arbitrary restrictions on current energy sources, rather than changing our infrastructure to use new energy sources, appears to be the aim of the Oregon Legislature.

8.2 Hydrogen

Hydrogen is an excellent fuel, producing water when burned and providing roughly two-thirds of the mileage of gasoline per gallon. However, it requires large amounts of electricity to hydrolyze water, so that without an inexpensive source of electricity, it's not practical to make hydrogen. Breeder reactors could provide this electricity, but President Carter severely set back the nuclear industry when he killed the reprocessing of waste back into nuclear fuel, creating our current waste crisis. Hydrogen has the advantage that in a collision, the gas escapes so rapidly that it doesn't explode like gasoline. This is also a disadvantage, though, as the smallest leak leaves a tank empty almost instantly.

Author's note: I can see driving along on of our wonderfully constructed Oregon streets and hitting a chuckhole or bump that breaks a poorly done weld on a fuel line. There's a loud, very short PPFFTTT! as the hydrogen molecules zip from the hole, moving at more than 50,000 miles per hour into the atmosphere. The hydrogen tank is emptied in a few fractions of a second— much faster than the fuel needle can move from Full to Empty. I am now stranded until the tow truck gets there. At least there is no puddle to clean up...

One has to note that hydrogen was not the cause of the Hindenburg explosion. Several outfits, including the Zeppelin Line's engineers, came to the conclusion that the coating on the fabric skin caught fire. The hydrogen, which burns blue, escaped almost instantly, while the fabric burned for about a minute.

8.3 Propane

Propane, otherwise known as liquefied petroleum gas, is a good substitute for gasoline. Its price hasn't changed much in recent years, staying around $2 per gallon. Propane is one of the gases naturally occurring in natural gas, as are butane and methane. It burns clean and is not as corrosive as gasoline, meaning engine seals and parts last longer. It has the disadvantage that the tanks are much more cumbersome than gasoline tanks. Propane is easily obtained and is available in most places in the nation. Propane doesn't produce particulates or gases other than water vapor and carbon dioxide.

8.4 Natural Gas

Natural gas is mostly methane, but contains fractions of propane, butane, ethane, and other hydrocarbons. The tanks required are similar to those of propane. It has the advantages of propane, but provides only two-thirds of the energy of gasoline per gallon. It is also much more explosive than gasoline or propane, making it more hazardous in a car wreck than gasoline or propane.

The cost of natural gas varies greatly around the U.S.. Some areas have very high prices due to supply problems—entirely because of politics. For example, California uses tremendous amounts of natural gas, but won't build LNG terminals to bring more in from Alaska, Canada or from Australia. New gas pipelines from Alaska and Canada are also a foregone idea, despite the need for them. Oregon has high prices because California uses most of the existing supply and Oregon won't use gas from fields that could be developed elsewhere in the state. Other areas of the U.S. have multiple sources of supply and much lower prices.

8.5 Methanol

Methanol, or "wood alcohol", is obtained as a byproduct from coal and other industrial processes. It's a poor fuel for vehicles. It can cause blindness both by drinking it or by inhaling it. As such, it is potentially hazardous to a driver in the event of leakage. It is extremely corrosive to engine parts and one has to fill up the tank four times as often as gasoline. Methanol will damage paint and many other coatings if spilled on them.

8.6 Ethanol

Ethanol is a fad that's ready to crash and burn. Unlike other liquid fuels, producing ethanol requires much more energy than it yields. First, diesel and gasoline are burned by the farm machinery used to till the soil, plant the seeds, and then in tending the field as the grain grows. Fossil fuel-based pesticides and fertilizers may be applied to the grain, if necessary. Fossil fuels are used by the harvesting equipment. Diesel is used by the trains and trucks that transport the grain to the ethanol plant.

Electricity, which is produced mostly from coal and natural gas, is used to grind up the grain and mix it with water and yeast. Electricity is used to keep the "mash" at a constant warm temperature during fermentation and then electricity is used to heat the mash in order to distill the alcohol from it. A whopping **ninety-eight percent** of the energy originally in the grain is used up by the yeast during the fermentation stage and the yeast excrete a mere **2%** of that original energy as alcohol. **Two percent efficiency!** That alone should have doomed ethanol as an alternative fuel. Fossil fuels are then used to transport the ethanol to processors and markets.

Ethanol has slightly over one half of the energy of gasoline, is highly corrosive and has to be "denatured" to prevent human consumption. Ethanol mixes are highly inefficient— adding ten percent ethanol to gasoline reduces the mileage per gallon by thirty percent. In cities like Denver and Portland that use a ten percent ethanol, ninety percent gasoline blend, 20% more gasoline is burned to make up for the reduction in mileage, which effectively trades one air pollution problem for another.

Ethanol production indirectly reduces much of the grain that would otherwise be used to make bread and noodles. The corn goes to ethanol and so more wheat is used for items that would have been made from corn and its derivatives. Farmers convert to corn growing from wheat and other grains because of the profit in corn. Much of the corn that was exported to China and Russia is now being converted into ethanol. That plus the reduction in the production of wheat and other grains has reduced overall grain exports. As a result, the price for food has risen worldwide.

Because of its inefficiency, ethanol would require immense acreage to power all United States vehicles. And since ethanol uses more energy than it provides, well, you can see the problem…

It would literally be more efficient to take the corn plants (roots, stalks, ears and all) now used for ethanol and burn them to produce electricity to run electric cars, even with the 50% lack of efficiency of the batteries in battery-powered vehicles.

Author's note: What is interesting is that twenty years ago (and even up to a few years ago), when much of the original research on alternative fuels was done by various outfits, the reference books stated that an engine using ethanol would get 1/3 the mileage of gasoline in actual use. Recently, it has been stated that it is 2/3 the energy of gasoline, leading people to believe that they will get 2/3rds of the mileage of gasoline if they use 100% ethanol. I have no confirmed explanation for this discrepancy.

I started to speculate that it may be the difference between the calories released when ethanol is burned in a calorimeter versus the actual mileage obtained (1/3rd) by a fuel that is notoriously inefficient in a combustion engine due to low combustion temperatures, but that doesn't work, either, because ethanol has slightly more than half the calories of gasoline per gram when burned. I've not found an official explanation, so I really don't know. I think it's probably "revisionist chemistry" to make ethanol sound like a better fuel than it actually is. For many years, it was stated by those who had converted to ethanol that it was 3 tanks of ethanol for each tank of gasoline when running pure ethanol.

8.7 Biodiesel

Biodiesel is a good fuel, but it can only replace diesel— biodiesel cannot be used in gasoline engines. (Perhaps someone can come up with a gasoline made from vegetable oils…) It burns cleaner than diesel and is roughly equivalent to diesel in performance. If oilcrops such as soybeans were grown for biodiesel, it would require much less acreage that ethanol-yielding crops would require (estimates vary). In addition, it requires much less energy to produce than ethanol as it does not require fermentation. Since it is not fermented, biodiesel doesn't have the 98% loss of the original energy. Were the U.S. vehicle fleet replaced with diesels, biodiesel would be a practical replacement for diesel made from petroleum, requiring only small changes to the engines. However, it would still require tremendous acreages of oil-bearing crops to replace diesel.

9.0 Chapter 9
Alternative Electrical Generation

There are a number of alternative methods for generating electricity besides turning coal and gas into electricity, which account for 49% and 20% of our electrical generation, respectively. Some are more efficient and cost effective than others.

9.1 Geothermal Energy

Geothermal energy is heat, usually in the form of hot water or steam, from springs or wells, that can be used to heat buildings or generate electricity.

Geothermal power has the advantage of using energy that already exists— nothing is burned and no waste is produced. Geothermal energy does not pollute the air, water or soil.

Most of the geothermal energy potential in the United States has not been developed. According to the Department of Energy and other sources, there is great potential for geothermal power in the western half of the U.S.. California, Washington, Oregon, Idaho, Nevada, Montana, Utah, Arizona, New Mexico, Wyoming, and Colorado have areas of high potential for geothermal power. If the USGS is correct, perhaps a million megawatts have yet to developed in this nation. That is the equivalent of 1000 hydroelectric dams the size of Bonneville Dam.

There are thousands of thermal springs in Oregon and California alone. Some localities are using their available geothermal energy, such as Klamath Falls, Oregon, which heats its city's entire downtown and many of the town's buildings with geothermal heat. The nearby Oregon Institute of Technology is completely heated by geothermal energy. Susanville, California heats more than 30 of its buildings with geothermal heat. Sonoma and Lake counties of California have "The Geysers", which are natural steam vents that are coupled to turbines, producing 850 megawatts of electricity with more being constructed.

There are farms west of Tule Lake, California, with well water so hot that the farmers have to let it cool before their cattle can drink it. Klamath, Lake, Malheur, and Harney Counties of Oregon each have hundreds of hot springs that vary in temperature from 70 to 180 degrees Fahrenheit. This kind of heat energy could be put to use as electricity.

Of 2000+ megawatts of geothermal electricity generated in the United States, California produces approximately 1800 megawatts of electricity with its 33 geothermal plants. This represents 5% of California's electrical generating capacity. Nevada is second to California with 15 geothermal electric power plants that generate 244 megawatts (2007 figures). There are a handful of plants elsewhere in the U.S., but California and Nevada account for most of the U.S. total electrical generation from geothermal energy.

2000+ megawatts sounds like a lot of electricity and it is, but it should be kept in mind that two nuclear reactors or a large hydroelectric dam like Hoover on the Colorado will generate the same amount of power. While geothermal is more expensive per watt generated than hydro or nuclear, it does have the advantage of decentralization, i.e., many small plants located in multiple locations, rather than one or two large units.

Since plants invariably are shut down at some point for maintenance or equipment failure, decentralization helps in maintaining constant power overall in the national electric distribution system. However, additional geothermal power generation in many new locations would require that improvements to the U.S. electrical infrastructure. The system of "interties" connecting the different electrical grids of the United States has been somewhat ignored since 1992.

It is also becoming almost impossible to place new major powerlines and even geothermal plants as environmental groups object to "visual pollution" or other petty reasons. In effect, such groups want electrical power for their homes, but not the infrastructure needed to create and deliver it. Several geothermal projects have already been delayed or ended by the environmental lobby's activism.

Adding several hundred megawatts of geothermal electrical generation would require new transmission lines and the upgrading of existing power transmission systems. Look at how an act of terrorism caused the electrical control systems to fail in seventeen states and parts of Canada in 2003. Imagine the stress to the existing system of adding a couple of thousand megawatt-plus power plants, which would be mostly in areas that currently have no transmission lines or limited ability to transmit power via lines. Transmission capacity has been improved in the west to feed California's ravenous appetite for electricity— made much worse by California's adamant refusal to meet its own electrical requirements— but most parts of the nation do not have this ability to efficiently transfer power from one region's grid to another.

To see where geothermal plants are in operation, see Oregon Tech's website: http://geoheat.oit.edu/dusys.htm One of my alma maters, Oregon Tech, maintains the Geo Heat Center for Oregon.

9.2 Nuclear

It may seem strange to call nuclear power an alternative electricity source, but since a reactor hasn't been built for at least 15 years and no modern designs have been put to use in U.S. commercial plants, "alternative" is probably appropriate.

France, by comparison, took several reactor designs, standardized them, and now 77.1% of their electricity comes from nuclear power and France exports almost 70 billion kilowatt-hours to neighboring countries. Three quarters of France's electrical generation burns no fossil fuels. It also produces no carbon dioxide, but the environmental groups always find a way to kick France's nuclear power, despite France having the cleanest electrical generation on average of any industrial nation, in terms of pollution produced per megawatt·hour. Indeed, if they truly believed that carbon dioxide causes Global Warming, then they would advocate nuclear power.

Nuclear power could potentially supply all of our electrical needs as well as provide the electricity to convert water into hydrogen to replace gasoline combustion in our vehicles. Despite the roach dreams of the left, NO other power source in the next thirty years could foreseeably provide the electricity to split water to create hydrogen at an economic price. If we want hydrogen gas-powered cars that cost less than $100,000 a year to run, we have to go nuclear. Despite this and despite being the *only* power source than doesn't produce *any* pollution, nuclear power has been almost driven to extinction by the environmental lobby— for years, lawsuits have been the most expensive part of a nuclear plant's operating costs— and by the mismanagement of nuclear waste.

The nuclear waste problem was initially created when President Jimmy Carter stopped the reprocessing of waste into new fuel rods. No Congress since then has seen fit to restart the reprocessing of the nation's nuclear waste, preferring to ignore waste entirely or to advocate stockpiling it in places like Yucca Mountain. While Yucca Mountain is as secure a location as any to be found for nuclear waste— the ground water is below 5000 feet of solid rock— it would be much better to recycle the waste through reprocessing and put it to use in generating electricity.

Nuclear reactors produce electricity without creating pollution, the "dry" designs recycle their cooling water and horror movies to the contrary, do not create havoc unless one has a criminal situation like Chernobyl. Chernobyl, which ironically means "wormwood" or "absinthe" in Ukrainian, was a poorly-designed plant tossed together by the Soviets primarily to produce plutonium. The badly maintained reactor was undergoing some tests of backup systems when the graphite-moderated reactor went critical.

When the fire began to burn up the graphite and other materials— an impossibility with U.S. and European plants— there was no containment dome to prevent the plume of radioactive materials from sweeping over thousands of square miles. Hundreds died, thousands were sickened, and more than 300,000 people had to evacuate and eventually move completely out of the area. Even with all of that, only a handful of places affected by the fallout have high levels of radioactivity today as the isotopes released by Chernobyl decayed quickly.

Compare that with Three Mile Island, where cooling water was deliberately shut down to the reactor— sabotage, as it was impossible to have happened by accident— and the worst thing that could happen to a U.S. reactor did happen. The core melted down. Despite that, except for pocketbooks and hysteria, not one person was hurt— no sickness or death from radiation as almost none was released. Long term studies show no increased cancer rates. Even if the reactor had somehow "run away" and undergone unrestrained fission, the meter-thick cast iron containment dome required of all nuclear plants in the west and the extremely strong, dense concrete would have prevented any escape of radioactive materials. Besides, there are limits to the amount of nuclear fallout that could be created due to the amount of radioactive material on location. The reactor would have run out of reactants long before the containment dome could be breached. The problem would be contained and eventually would be cleaned up, with no loss of radioactive materials.

Author's note: Try using a logical argument like that last with the environmentalists, who get their information about nuclear power from Hollywood, and from their fellow Econuts who know less than they do and from cannabis sativa cigarettes. A logical argument is wasted on those who don't care about facts, figures, physics or chemistry. The late mass murderer and dictator Vladimir Illych Lenin called people like that "useful idiots", because they were enthusiastic about the Communist cause but didn't understand what Communism was all about and when their usefulness was ended, Lenin killed or exiled them to Siberia..

Breeder reactors, or fast fission reactors, as they are sometimes called, while now in use in France, have not been adopted in the United States, due primarily to the efforts of the environmental lobby. Breeder reactors produce roughly 1% of the waste of our current commercial reactors, and have greater safety features than any reactor currently used for electrical generation in the U.S.. Most require less maintenance as well. A number of designs exist, but until the current crop of environmentalists decide that it's better to have 200 new nuclear power plants that produce no pollution, than 600 or so current coal-fired plants, there will be no new nuclear reactors. Author's note: Thank you ever so much, environmentalists, for limiting us to coal power, the dirtiest form of electrical generation that we have through, your tireless efforts to stop anything new or more efficient.

9.3 Fusion

Fusion is a power source that holds promise for the future, but is not in a state of development that will allow its use any time soon. The basic idea is that hydrogen isotopes like deuterium and tritium are heated and compressed to 100 million+ degrees and the hydrogen atoms merge, forming helium atoms. There are two main methods for doing this: magnetic containment and inertial containment. Both are being researched.

The technical problems of heating hydrogen gas to 100 million degrees and containing the hot gas have been daunting, much less keeping the atoms in close proximity long enough to fuse. Despite fusion's reputation, it's much dirtier than fission, bombarding reactor machinery with neutrons, making them radioactive. Several reactor designs show promise, but we still are having problems initiating fusion, much less reaching the "break-even" point, where we generate as much energy as is used to start fusion and keep the reaction going.

Cold fusion is fascinating, but Japan and France have been unable to consistently produce ignition and keep a constant reaction going, despite spending hundreds of millions on the process. While it's easy enough to get a few neutrons out of a cold fusion experiment, so far it's unreliable as a power source. If cold fusion is perfected, or at least becomes commercially useable, it could solve our energy problems. For now, though, it's a scientific curiosity.

9.4 Solar

I've always liked the idea of getting electricity from the sun. Solar energy is very good at producing small amounts of electricity in locations where sunlight is abundant. There are two fundamental problems with solar power that need to be solved— it's expensive, ranging from $3.50 to $10.00 per watt of electricity that a solar panel can produce and it's low efficiency (a maximum of 14% of the sun's energy striking the solar cells can be converted as of this writing), meaning that solar cells have to have large surface areas to generate much electricity.

Roughly 120 watts of energy of different wavelengths are present in a square foot of full sunlight, but with a typical efficiency of 12%, that means only about 14 watts are generated per square foot of solar panel under ideal conditions which means that you will get less than that most of the time. Call it nine or ten watts typical output.

For a house that requires 10,000 watts on average, that means solar panels of 1000 square feet (31.6 feet by 31.6 feet, if square) at a cost of $3,500 to $10,000 dollars for power during daylight hours only (to provide enough for power to store for other times would require much more than that). New types of solar cells are constantly being developed, but I've lost track of the types that showed promise, but never actually hit the market. Eventually, solar will begin to replace some of our current power sources as the problems are solved, but that may be some decades down the line at the rate things are progressing.

The most efficient home setup for solar in terms of cost is as a supplementary power source without batteries. In places like California or Arizona that receive many days during the year of sunlight, solar cells can be used to replace some of the power used in a home during the day. Grid power would make up what the solar panels didn't supply and would supply power during the night. Peak hours for electrical utilities are during the day and if many homes did this, it would reduce the peak demand that the utilities had to meet, saving money for the customers in the long run in reducing the need for more power plant capacity during peak hours.

Some homes use solar exclusively, but this means a much larger solar array, one that can not only meet the power needs during the day, but is large enough to store several times the energy required during a given day away in batteries. Batteries typically require double the energy placed into them for what one removes later and they generally last less than five years, depending upon the model, adding cost and creating the problem of what to do with the defunct batteries. Lead is toxic and lead-acid batteries are still the best for such applications, despite first being created a hundred plus years ago.

9.5 Wind

Wind power is useful and can generate considerable power, but it is dependent upon consistent moderate to high velocity winds. It's much cheaper than solar power, and in areas like the Columbia Gorge, Mojave Desert, or the Great Plains that receive a lot of wind, it's practical. Some environmentalists don't like it due to the occasional bird casualty, but it does not pollute and while it is usually noisy, wind power has great potential now that many of the technical problems have been solved.

Were the U.S. to replace our current electrical generation with wind power, it would require at least 3 to 4 times the generation capacity of the current U.S. electrical demand, spread out all over the U.S. in rural areas, linked with a much-upgraded system of transmission lines to transfer power from areas with high winds to areas where winds are not blowing enough at the moment.

9.6 What Should the United States do?

Everyone has their own ideas, but if it were somehow up to me to plan our energy use, this is what I'd want to do: Over a period of ten years or so, I'd replace all current electrical generation with breeder reactors. I'd build enough electrical-generating capacity to be able to produce and distribute hydrogen to replace fossil fuel use in our vehicles. I'd reprocess all current nuclear wastes into usable fuels for the new reactors and convert the small amount of leftover waste into materials for nuclear medicine. The U.S. has greater proven uranium reserves than any other nation, has the technology and the resources, so this would be reasonable for us to do.

While costly to do the initial conversion, this scheme would free us from the tyranny of the oil-producing nations. It's ludicrous to import energy from the Arabs in the Middle East and national socialist Venezuela. There would finally be stability in our energy supply. It would virtually eliminate smog from our cities. The sulfur compounds from our vehicle exhaust and coal-electric power plants would no longer be produced, eliminating the remaining acid rain in the eastern U.S. and most of the smog. If the nuclear power industry was properly regulated and protected from frivolous lawsuits, electricity would be less expensive in the long run than any source except hydropower. Inexpensive energy would stimulate the economy.

But we'll not change our energy production until oil is $200 a barrel or there is another Arab-made energy crisis like 1973 or 1979. There are too many interests vested in the status quo, as well as the environmental lobby's illogical and puerile hatred of anything nuclear.

People like Senator Gore and Speaker of the House Pelosi resist the idea of nuclear power, despite the incredible benefits to the environment of abundant and inexpensive electricity with no pollution. Instead, they'll pursue their own socialist agendas at the expense of our nation and its people.

It's what they do.

10.0 Chapter 10
Conservation Decision-Making

Making good decisions seems to be a rapidly declining skill in government circles in the last 40 years. To my unphilanoitosic viewpoint, wisdom is the ability to foresee the results of decisions. By that definition, wisdom is almost gone amongst our ranks of politicians and our government civil servants. Unfortunately, as long as the American public forces politicians to lie to them, and tends to ignore those politicians who don't lie to them, wisdom in government will continue to decline. The ability to make sound decisions is crucial to our future. Here are some ideas for the land manager who is being buffeted by management fads, the environmental lobby and by land users.

10.1 Think Things Through

This may seem like a basic thing for making decisions, but the environmental lobby very rarely thinks matters through when they decide to protest or launch a lawsuit. As a general rule, they latch onto an idea and only later notice the downsides. Think matters through. A few minutes of reflection can save a lot of wasted hours.

For example, electric cars have been pushed by the environmental lobby. No pollution, right? Wrong. One of the major problems with electric cars is that the power to charge the batteries has to be generated elsewhere, mostly through coal power generation. The electricity is transmitted via power lines and then is used to charge the batteries. Some power is lost in transmission, more is lost in charging the batteries, so that only half of the originally generated electricity is available for the car to use. Ultimately, the amount of waste products from combustion to produce the energy is effectively doubled over that of a gas- or diesel-powered car.

Further, batteries are either lead-acid, or nickel-metal hydride (possibly lithium ion in the future) and these require toxic materials. The batteries have to be replaced after a few years of use and so present a problem with disposal/ recycling the old batteries. In the event of a wreck, electrolyte from the batteries can be spilled, causing harm to the car's occupants or others nearby and presents a chemical hazard that must be cleaned up. The electricity stored in the batteries can be a shock hazard in an accident. The batteries can be shorted and some battery types such as lead acid and (especially) lithium ion can explode. Rarely are these problems mentioned by those that advocate for electric cars.

Hybrids were initially disliked by the environmental lobby— partially, because they still used diesel or gasoline, but mainly because the idea didn't originate with them. Hybrids average about 10 miles per gallon more than their non-hybrid counterparts, and so some of the environmental lobby now pushes them as the current way to get to heaven. The hybrids have the problems of both fossil fuel engines as well as the problems of batteries mentioned above, plus the technical problems of making it all work. By the time one mines the nickel and iron and creates the chemicals for the batteries, which have a limited lifetime, are electric cars and hybrids really saving anything, in environmental terms? Probably not.

Another case: feral cats are often demonized. Most cities and towns go to great lengths to remove them from the streets and they usually end up killed at animal shelters. But consider for a moment where feral cats live and what they eat. They eat mice and other rodents in large numbers! I used to believe that it was awful to have feral cats in towns. Here in the Pacific Northwest, there aren't many feral cats except in the temperate areas west of the Cascades because the most areas on the east side of the Cascades typically have 180 to 240 days below freezing, which is too cold for feral cats to survive without human help. It was never much of an issue.

Then, when I worked in semi-rural northern California, I ran into feral cats and thought it awful. Initially, I agreed with the usual logic of getting them off the streets. (I love animals and hated to see them on the street. Still do. Wish there was a better way to get rid of rodents). Then, two studies came out in the early 2000s that indicated high levels of hantevirus in the local mice. The only effective defense one has against rodents in towns and cities are feral cats— our well-fed house cats generally are pretty lazy. Suddenly, it was "nice kitty: go eat those mice", when I saw one around the neighborhood.

Portland, Oregon has waged a war against feral cats in recent years. Now, they are being overrun with the black rat and the Norway rat. While hantevirus is not very common yet in the north end of the Willamette Valley, there are other diseases that rats carry. Rats are showing up in people's homes, sheds, cars, etc.. I still don't like the idea of feral cats, but there aren't any alternatives to cats other than putting poisons out on every street and alley. Imagine the problems with doing that. Kids and pets possibly getting into the poison, rain washing it away, snow, etc.

The above are examples of how not to make decisions, because the conventional ideas led to worse problems. Think things through. Don't make snap decisions, either. Except obvious emergencies like fires, chemical spills and the like, most environmental issues don't require an immediate judgement and sometimes matters are clearer after a few days of researching the matter. Take everything that is known about an issue into account before determining what the best option is.

For example, on the feral cat problem, perhaps it would be a better idea to control the populations of feral cats by spay/neutering, give them the regular immunizations for cats, and treat them more like public assets for vector control, rather than to remove them outright. Another possibility for northern California and southern Oregon would be to breed the local ringtails (miner's cats) and find ways to keep them in town to eat the mice. They are often friendly to humans (they would domesticate miners, and hence the name) and are more effective at mousing than cats. There would be many issues to sift through with a proposal like that, however.

10.2 Ignore Fads

Like most fields of human endeavor, fads are rampant in the environmental and natural resource community. Ignore them. Right now, Global Warming and SUVs are the "evil" things that must be stamped out. As has been said before, Global Warming due to man's actions is a myth at best. SUVs are used because families need a larger vehicle, both for safety— per 1000 accidents, one is twice as likely to die in a small car than an SUV— and to be able to haul their kids around, much as the station wagon used to be used. In many areas, especially California and Oregon, SUVs are under attack because of the anti-SUV fad. That half again the cost in fuels and materials would be required if a family bought two small cars to move about instead of buying an SUV eludes those who advocate small cars versus "evil" SUVs.

Global warming is perhaps the worst current fad. Radio, TV, newspapers, environmental groups, politicians are constantly yammering about "Climate Change" and "Global Warming." If it is warmer or colder than average, wetter or dryer than average, or exactly average, it **must** be the product of Global Warming. That any change in the global climate is beyond any of man's current abilities and thus, not man's responsibility is completely ignored. Their reasoning is: the climate is changing. To fix it, let's destroy the western economies and reduce our populations to poverty. May it never be. Never make decisions based on fads. They never pan out over time and have sometimes caused great sorrow.

10.3 Understand the Mechanisms Involved

It's unfortunate, but most people who live in urban areas do not understand how nature works, how farming works, how ranching works and indeed, how the natural world in general works. Most environmentalists come from cities, and they, too, generally don't understand how natural systems work. That is responsible, at least in part, for many of the ideas that environmentalists have and the misconceptions that they labor under.

To make a good conservation decision, one has to understand how things work. This seems basic, but, again, most people don't have the background or experience to understand how natural systems function. The decisions involving conservation have often been characterized by ignorance during the past decade or so on the West Coast because of this lack comprehension by special interest groups and even some land managers.

For example: I remember being surprised by some Forest Service range managers who said some land was in good shape one day when I accompanied them to a mountain cattle allotment. The ground was compacted badly, the ground pockmarked and the grass had been eaten so often that there was more bare sand and dirt than grass left. What grass was present was sparse and low-cut. The nearby stream had been partly denuded, the soil a mass of pocks and the vegetation mostly gone. On even a quick glance, I would have said it was overgrazed and been for several years.

At this point, the land needed some serious "time off" to recuperate and some restorative work. The problem is: what about the farmer/rancher? The answer might be to get a conservation grant to do the restoration work and tide the farmer/rancher over for a couple of years for the land to recover.

With some restorative work and timing, ranchers in the Prineville, Oregon area, for example, found that they could run four times as many cattle on an allotment. That's serious money for a cattle rancher. This involved the timing of placing the cattle on the allotment, fencing the riparian area— except for a few spots for watering the cows— and the restoring/raising of the streambed so that the water table near the stream rose to its former level again, which in turn caused the floodplain to green up again, which provided a LOT of extra food for the cattle as well as benefits to local wildlife and plants. The repair of the riparian area increased streamflows during the dry portion of the year as well, serving as a regulator for the water. More work, but a lot more money for the rancher. More fish, too, for the fisherman because of the stream conditions.

The current fad among environmental groups is to prevent logging and forest maintenance of any sort in second growth forest stands. This shows a complete lack of understanding of how a forest works. As mentioned in Sections 2.5 and 5.7, a second growth forest requires maintenance to prevent catastrophic fire. Since the environmental lobby has determined that all forest management is somehow bad or unnatural, they have fought any sort of maintenance tooth and nail. This has helped create the recent catastrophic fires we have seen— many areas could have been maintained and both the intensity and chance for fire in those areas could have been reduced if environmental groups hadn't filed lawsuits. They don't understand the mechanisms of nature, even as they claim to "defend" nature.

10.4 Just the Facts

Emotion causes a lot of terrible decisions. Indeed, rampant emotion replacing logic and fact is one of the fatal flaws of the environmental movement as well as our Federal, state and local governments. I've lost track of the times I've seen this scenario: an environmentalist picks up a rumor, asks a friend or two about it— doesn't take time to research the facts— and then goes on a blitz to "save the scarlet dirtfish" or something similar. Often, the decisions ultimately made in the face of such pressure have lasting negative consequences.

For example, a decade ago, animal-rights organizations decided to get rid of cougar hunting in Oregon. They sponsored an illegal hunt where a hunter had his dogs tree a cougar and then had the dogs attack the cougar. Totally illegal, and hunters just didn't do that sort of thing. "Why waste a good cougar?" and "Why would I want to injure my beloved dogs?" is what hunters would think. The fact that setting up such an illegal hunt would cause a cougar and the dogs to get mauled to death apparently was okay as far as the animal-rights groups were concerned, presumably because it was for the "greater good" of getting rid of cougar hunting.

The animal-rights groups put the fake hunt on a TV commercial for a ballot measure that would outlaw the use of dogs to track cougars. They misrepresented the hunt video as being just an ordinary occurrence. The state's voters stupidly voted to ban dogs for tracking, getting up in arms about the perceived horrible things done to cougars by the awful hunters. Within a decade, cougar numbers had went from ~5000 to ~9000 (2006 Oregon Department of Fish and Wildlife statistics) and deer populations had dropped.

Worse, with cougars invading urban areas in search of new territories, there is greater possibility of cougar-human interactions than before. Attacks are rare, but menacing by cougars is not uncommon. Cougars require a lot of territory and when the population exploded, a lot of young cougars were pushed into the cities, as most of the existing habitat was already occupied by cougars.

In the Bend area, several problem cougars have been removed in the last three years— one of them 3 blocks from where I live. The cougar had been living under a porch and for a couple of weeks the local dogs and cats wouldn't go near the house. Eventually, someone noticed the cougar. Cougars are generally not dangerous to humans, but children are at greater risk, as are household pets. In October 2007, cougar signs were found in a neighborhood less than a quarter of a mile from my home. Cougars were a rare sight prior to the measure being passed. In recent years, I've seen cougars in our forests more often that bears or bobcats and a decade ago, one didn't see cougars as a general rule.

Emotion caused the cougar measure to pass. The facts at the time and especially in hindsight, argued against the measure passing. This is just one example of emotion resulting in bad decisions. Get the facts.

11.0 Data Sources

All data sources used in the preparation of this book are in the public domain, i.e., anyone can look them up, check them out and use the information.

Climactic data was derived from the databases of National Aeronautics and Space Administration (NASA), National Oceanic and Atmospheric Administration (NOAA), U.S. Geological Survey (USGS) and several other federal agencies.

Sunspot and Solar Irradiance data was derived from the National Solar Observatory and NASA databases.

Atmospheric data was derived from NOAA databases.

Chemical and thermal properties of atmospheric gases were derived from common scientific, public domain sources.

Historic information was derived from the archives of the National Weather Service.

Forestry data was derived from the archives of the U.S. Forest Service and USGS.

12.0 Scripture References

All scripture references contained in this book are from the Authorized Version of 1611, otherwise known as the King James Version.

13.0 Glossary

Board Foot   A unit of lumber, defined as 1 foot by 1 foot by 1 inch.

Environmental Lobby The environmentalists, their organizations and their politicians, such as Gore and Pelosi.

Good for the Planet Whatever the environmentalist or his/her group says is good for the planet. Often arbitrary or fad-driven.

Green   A somewhat nebulous term meaning something is not harmful or is beneficial to the environment. Usually a product of fad, rather than fact.

Natural   This term is applied to pretty much everything that is not man-made. It's also pretty nebulous, being without an exact meaning.

Progressive   Anything socialist. Taking private property for public use is progressive. Taking from the rich and giving to the poor is progressive. Taking from the poor and giving to the rich is progressive. Taking from the middle class and giving to both rich and poor is progressive.

Social Justice   An arbitrary measure that says that some people in society should have more than others in terms of public aid, to bring them up to a level of material wealth. The problem with this is that it creates a permanent underclass— as we saw with Lindon Johnson's "Great Society"— and destroys the family. Politicians push this to gain votes, despite the deleterious effect on those so "helped" and the public in general.

Spin   A term the media and politicians use. It means to creatively lie, change facts or at least make facts more palatable to the public in some way.

Philanoitoser   One who loves foolishness. To create the word, I took the Greek roots philo, which is love for something and anoitos, which is foolishness.

14.0 Glossary of Federal Agencies

BLM Bureau of Land Management. This agency deals primarily with managing publicly-owned grasslands and desert/steppe lands of the central and western United States and forestlands in Oregon and northern California.

NASA National Aeronautics and Space Administration. This agency deals with anything to do with space and the upper atmosphere. Satellites, spacecraft, astronautics, astrophysics and such.

NOAA National Oceanic and Atmospheric Administration. This agency deals with studies of the ocean and atmosphere, overlapping somewhat with NASA. It maintains buoy networks and weather satellites, oversees the NWS and the National Marine Fisheries Service.

NSO National Solar Observatory. This observatory is a division of the National Science Foundation that studies solar phenomena.

NWS National Weather Service. An agency under NOAA, this agency studies weather and provides forecasts and weather alerts.

USFS U.S. Forest Service. An agency under the Agriculture Department, this agency is tasked with maintaining most of the federal forestlands and some National Grasslands. They have done extensive research on forestry over the years and maintain data centers around the nation.

USGS U.S. Geological Survey. This agency should really be called the U.S. Cartographic Agency, as it maintains mapping databases for plants, animals, land use, forests, populations, geology, hydrology, agriculture and the National Atlas. It also maintains irrigation services in some parts of the west and studies volcanism, earthquakes and other natural phenomena.

15.0 Appendix 1 Socialism

There are different branches of socialism including communism, fascism/national socialism, modern international socialism and parliamentary socialism. These systems differ in their levels of economic control and the freedom that their citizens (or subjects) retain. All socialist governments, regardless of type, have two general social classes— the elite, who have the power, money, etc., and everyone else.

In pushing socialism, progressive politicians and the environmental organizations plan to be among the elite group. Sadly, people like that were called "useful idiots" by Vladimir Ilych Lenin, who used such people to further his ends and then killed them or exiled them after their usefulness was ended. If the Global Warming advocates succeed in bringing world socialism, most of them would not fare well once a socialist revolution succeeded, if history is any guide.

Socialist governments in Europe and Asia resulted in the deaths of 110 to 250 million people during the 20th Century from warfare, genocide and famine. World War 2 in Europe killed as many as 50 million. Dictator Josef Dzhugashvilli (Stalin) of the Soviet Union killed more than 30 million of his own citizens prior to World War 2. During the Great Patriotic War—the Russian name for WWII—13 million Soviet communists died fighting the German national socialists due to the commissar system in 1941-42 and later, the Soviet practice of using "undesirables" in "shock armies". Perhaps ten million civilians each in Russian and Germany were killed by the armies as they moved first east into the Soviet Union and the Ukraine and then west into Poland and Germany. Another ten to twelve million civilians were killed in Hitler's death camps, including six million Jews.

In China, the Chinese Civil War, the Great Leap Forward and the Cultural Revolution each killed roughly 30 million apiece. Mao starved entire cities to death when it suited him during the Chinese Civil War. The Khmer Rouge of Cambodia murdered 2 million of their "intelligentsia". The North Vietnamese killed off at least 3 million in their war with South Vietnam and the U.S. and in "purges" after the war was over. Other socialist nations had fewer casualties, but ruthlessness by the ruling elite was the norm.

16.0 Appendix 2 Types of Socialism

16.1 Communism

Communism is socialism where the entire economy of a nation is planned and executed by the government. (In more ways than one— communist governments are incredibly inefficient and tend to kill the economy they plan.) Examples of communism are the former East Bloc and the former Soviet Union. No current government is strictly communist, although several governments, such as that of China, still claim to be communist. Life has no value under communism and extreme poverty was typical of communist nations, with the exceptions of the former communist nations of East Germany, Poland and Czechoslovakia, where conditions were somewhat better.

Health care was largely ignored in the communist nations and basic items like food, clothing, soap, and medicines were often scarce. This government type tends to expand by warfare and threat of warfare. Law does not rule; merely the caprice of the "elite". Politics are often draconian, with political enemies exiled or killed.

16.2 Fascism/National Socialism

Fascism/national socialism is characterized by key industries controlled directly by the government. Private businesses and property rights are allowed for the most part. The government and politics are strictly governed by the ruling party and nationalism and propaganda are used to cause the populace to back the government. The World Workers Party (WWP), a communist party active in much of Europe, gave birth to this type of socialism after the First World War.

In Italy, Benito Musselini's Face of Socialism Party, from which we get the name "fascism", grew from the Italian branch of the WWP and in 1919, the German Workers Party, the German branch of the WWP, was renamed the National Sozialist (Nazi) Party. Germany of 1933-1945, Italy of the 1920s-1945, Greece and Romania of the 1930s-45, were national socialist. China from 1995 or so to the present has been steadily changing toward national socialism, and Venezuela from the end of 2006 to the present has become a national socialist nation.

Violence and murder are used to neutralize or frighten opposition. Governments of this type often use scapegoats to take the blame for social and economic inequalities, frequently using propaganda to dehumanize opponents. Nazi Germany did this with the Jews and communists. China dehumanizes the West and Hugo Chavez the United States. Fascists/national socialists of the 1930s used the term "useless eaters" and desired the human race to be purged of them. Life has some value under fascism/national socialism, but this government type also expands by warfare and threat. The rule of law is for all but the scapegoats and Party members.

While often called the "right" in politics, I submit, that, as a branch of socialism, National Socialism is, by definition, left-wing.

16.3 Parliamentary Socialism

Parliamentary Socialism is a more moderate form of socialism mixed with democracy. Canada and most of Europe have this government form today. The rule of law is still in place for the most part, but it is eroding rapidly. Life had great value initially, but it too, is fading quickly. For example, euthanasia has become "popular" in nations like the Netherlands and is often perceived now as a duty to die, rather than a choice. Drug use and crime have become very common in the last decade and now Canadian and European culture and social cohesion are breaking down.

These nations were true parliamentary democracies prior to adopting socialist ideology in key areas of government during the mid-1900s. Originally touted as providing free medical care and other benefits to everyone, these nations' economies are now stagnating and mortality due to the lack of available health care is now severe. Most Canadians with serious illnesses now come to the United States for medical treatment because of delays in treatment or the unavailability of the same treatments in Canada. Canadians recently began coming to the U.S. to have their children, as well. Many Europeans now travel to other nations for health care or use expensive private clinics and hospitals when public health care is rationed, delayed too long or simply not available.

One fifth of all deaths in the U.K, for example, are the result of delayed care, according to a 2007 British study. Antibiotic-resistant staph (Methicillin Resistant Staphylococcus Aureus or MRSA) is now a serious problem in Europe and could be stopped by simple hospital cleanliness. It is only now becoming a problem in some of the United States' clinics and hospitals.

16.4 International Socialism

Modern international socialism is similar to fascism/national socialism, differing only in scope. International socialism is not national, as in the German Volk (people) of Nazi ideology, but rather the entire world is seen as "needing" socialism to survive. At least that is the excuse— the real reason for pushing international socialism is the desire for power. This government philosophy has become very popular in Europe, parts of the United States and Canada and has become more and more radical in recent years. Some of the United States' environmental groups, many of the modern Democrat Party leadership and so-called "anarchist" groups are international socialist in philosophy.

International socialism seeks a world government, disdains national boundaries and individuals, wanting the whole of humanity to embrace their ideals of socialism. Violence and protest are used as tools, while at the same time the leadership denounces violence, just as prior socialist movements used violence and protest while proclaiming innocence. International socialism uses the same methods of propaganda and scapegoating that fascism/national socialism does. Progressives demonize the center/right and Christians, much as the National Socialists demonized the Jews and dehumanized the Slavs.

The dehumanizing rhetoric supporting abortion used by progressives is illustrative of socialism in statements about the developing baby such as "it's not human", "it's better off dead", etc.. The same arguments were used by Hitler and the National Socialists in regards to the Jews in Europe from 1919-45. The National Socialists of Germany's Weimar Republic of 1930 represent the closest historic parallel to the current state of the international socialists in world politics. This is truly frightening.

With International Socialism, one has all of the evils of the Italian Fascists and German Nazis, but with one additional disadvantage— unlike the Nazis, the International Socialists have no loyalty to their own countries, or indeed, anyone but themselves, being completely selfish and self-absorbed.

Some progressives and environmentalists of this branch of socialism have begun advocating reducing the world population to 300 million and sometimes even 100 million, depending upon the individual asked. They usually digress on how this will be accomplished and when pressed, speak of methods that are truly evil, such as withholding food or medical aid to nations with large populations, or even deliberately spreading diseases to kill whole populations. That would be unspeakably criminal. It would be best for both our nation and the world if the international socialists didn't progress further. God willing, the world will see them for what they are and what they want to do.

17.0 Appendix 3 Thermal Effects of a Carbon Dioxide Increase on the Atmosphere

It requires an understanding of the physics of gas to understand the thermal aspects of how atmospheric gases absorb heat and change temperature. The pressure of one standard atmosphere at sealevel is defined as 1 bar or 1000 millibars. The amounts of the different atmospheric gases are described by a term called "partial pressure", which basically means that if oxygen is 21% of the atmosphere, then it's partial pressure accounts for 210 millibars of the total pressure of the 1000 millibars that make up standard atmospheric pressure.

The partial pressure of nitrogen in Earth's atmosphere is said to be 781 millibars. Each gas has a given "specific heat", which is the amount of heat energy required to raise the temperature of the gas 1 Celsius degree. So, to raise the temperature of one gram of hydrogen gas 10 degrees would take 143 Joules. A Joule is a unit of energy and is roughly equivalent to ¼ of a calorie.

Table 4 indicates what would happen if CO_2 doubled in terms of the changes to the thermal properties of our atmosphere. Table 4 does not include water vapor due to the increased complexity of the calculations (I strove to keep this book as easy to read as possible). In a real atmospheric system with water vapor, the doubling of CO_2 would have even less of an effect upon the atmosphere due to water's extremely high specific heat. Water vapor accounts for most of the heat that the atmosphere is able to absorb and release.

Table 4 shows that the total change caused by the doubling of carbon dioxide is a reduction of 0.04% in the amount of energy required to raise one gram of Earth's atmosphere one degree, not counting the effects of water vapor. This number is probably too large, given the accuracy of the numbers here, so doubling the carbon dioxide in our atmosphere does nothing in terms of heating or cooling. This alone is enough to dispel the notion of Global Warming as the result of man's actions.

Now, when water vapor is added to the equation, which would be a more realistic situation, the change becomes zero percent. A doubling of atmospheric carbon dioxide would have no effect on the thermal properties of the atmosphere.

Table 4   Effects of CO_2 Doubling on Heat Capacity of Dry Atmosphere at Sealevel

Gas	Chem Fmla.	Percent	Current Partial Pressure in Millibars	Specific He (Cp) at 70°F kJ/kgK	Partial Spec Heat	Pressure in Millibars After CO2 Doubles	Percent After CO2 Doubles	Partial Spec Heat After CO2 Doubles
Nitrogen	N_2	78.08%	780.84	1.0400	0.8120736	780.32	78.057529%	0.811537631
Oxygen	O_2	20.95%	209.47	0.9170	0.192084	209.33	20.939899%	0.191957215
Argon	Ar	0.93%	9.34	0.5200	0.0048568	9.33	0.933683%	0.004853595
Carbon Dioxide	CO_2	0.033%	0.33	0.8400	0.0002772	0.66	0.066021%	0.0005544
Neon	Ne	0.00182%	0.0182	1.0293	1.873E-05	0.018188	0.001819%	1.87202E-05
Helium	He	0.00052%	0.0052	5.1930	2.7E-05	0.005197	0.000520%	2.69858E-05
Methane	CH_4	0.0002%	0.002	2.4800	4.96E-06	0.001999	0.000200%	4.95673E-06
Krypton	Kr	0.00011%	0.0011	0.2480	2.728E-07	0.001099	0.000110%	2.7262E-07
Sulfur dioxide	SO_2	0.0001%	0.001	0.6400	6.4E-07	0.00099934	0.000100%	6.39578E-07
Hydrogen	H_2	0.00005%	0.0005	14.3200	7.16E-06	0.00049967	0.000050%	7.15527E-06
Nitrous Oxide	N_2O	0.00005%	0.0005	0.8800	4.4E-07	0.00049967	0.000050%	4.3971E-07
Xenon	Xe	0.000009%	0.00009	0.1580	1.422E-08	0.00008994	0.000009%	1.42106E-08
Ozone	O_3	0.000007%	0.00007	0.8200	5.74E-08	0.00006995	0.000007%	5.73621E-08
Nitrogen dioxide	NO_2	0.000002%	0.00002	0.8088	1.618E-08	0.00001999	0.000002%	1.61647E-08
Iodine	I_2	0.000001%	0.00001	0.2140	2.14E-09	0.00000999	0.000001%	2.13859E-09
				Total Spec			Total Spec Heat:	1.0089621
Assumes 70°F and 1000 millibars pressure				Heat:	1.0093509			
							Total Change:	0.04%

www.ingramcontent.com/pod-product-compliance
Lightning Source LLC
Chambersburg PA
CBHW081213170526
45165CB00009B/2810